Developments in Solid Oxide Fuel Cells and Lithium Ion Batteries

Technical Resources

Journal of the American Ceramic Society

www.ceramicjournal.org

With the highest impact factor of any ceramics-specific journal, the *Journal of the American Ceramic Society* is the world's leading source of published research in ceramics and related materials sciences.

Contents include ceramic processing science; electric and dielectic properties; mechanical, thermal and chemical properties; microstructure and phase equilibria; and much more.

Journal of the American Ceramic Society is abstracted/indexed in Chemical Abstracts, Ceramic Abstracts, Cambridge Scientific, ISI's Web of Science, Science Citation Index, Chemistry Citation Index, Materials Science Citation Index, Reaction Citation Index, Current Contents/ Physical, Chemical and Earth Sciences, Current Contents/Engineering, Computing and Technology, plus more.

View abstracts of all content from 1997 through the current issue at no charge at www.ceramicjournal.org. Subscribers receive full-text access to online content.

Published monthly in print and online. Annual subscription runs from January through December. ISSN 0002-7820

International Journal of Applied Ceramic Technology

www.ceramics.org/act

Launched in January 2004, *International Journal of Applied Ceramic Technology* is a must read for engineers, scientists,and companies using or exploring the use of engineered ceramics in product and commercial applications.

Led by an editorial board of experts from industry, government and universities, *International Journal of Applied Ceramic Technology* is a peer-reviewed publication that provides the latest information on fuel cells, nanotechnology, ceramic armor, thermal and environmental barrier coatings, functional materials, ceramic matrix composites, biomaterials, and other cutting-edge topics.

Go to www.ceramics.org/act to see the current issue's table of contents listing state-of-the-art coverage of important topics by internationally recognized leaders.

Published quarterly. Annual subscription runs from January through December.
ISSN 1546-542X

American Ceramic Society Bulletin

www.ceramicbulletin.org

The *American Ceramic Society Bulletin*, is a must-read publication devoted to current and emerging developments in materials, manufacturing processes, instrumentation, equipment, and systems impacting the global ceramics and glass industries.

The *Bulletin* is written primarily for key specifiers of products and services: researchers, engineers, other technical personnel and corporate managers involved in the research, development and manufacture of ceramic and glass products. Membership in The American Ceramic Society includes a subscription to the *Bulletin*, including online access.

Published monthly in print and online, the December issue includes the annual *ceramicSOURCE* company directory and buyer's guide. ISSN 0002-7812

Ceramic Engineering and Science Proceedings (CESP)

www.ceramics.org/cesp

Practical and effective solutions for manufacturing and processing issues are offered by industry experts. CESP includes five issues per year: Glass Problems, Whitewares & Materials, Advanced Ceramics and Composites, Porcelain Enamel. Annual subscription runs from January to December. ISSN 0196-6219

ACerS-NIST Phase Equilibria Diagrams CD-ROM Database Version 3.0

www.ceramics.org/phasecd

The ACerS-NIST Phase Equilibria Diagrams CD-ROM Database Version 3.0 contains more than 19,000 diagrams previously published in 20 phase volumes produced as part of the ACerS-NIST Phase Equilibria Diagrams Program: Volumes I through XIII; Annuals 91, 92 and 93; High Tc Superconductors I & II; Zirconium & Zirconia Systems; and Electronic Ceramics I. The CD-ROM includes full commentaries and interactive capabilities.

Developments in Solid Oxide Fuel Cells and Lithium Ion Batteries

Ceramic Transactions Volume 161

Proceedings of the 106th Annual Meeting of The American Ceramic Society, Indianapolis, Indiana, USA (2004)

Editors
Arumugam Manithiram
Prashant N. Kumta
S. K. Sundaram
Siu-Wai Chan

Published by
The American Ceramic Society
PO Box 6136
Westerville, Ohio 43086-6136
www.ceramics.org

For information on ordering titles published by The American Ceramic Society, or to request a publications catalog, please call 614-794-5890, or visit our website at www.ceramics.org

ISBN 1-57498-182-X

Contents

Lithium Ion Batteries

Preface

The growing environmental concerns and the increasing depletion of fossil fuels have created enormous global interest in alternative energy technologies. Fuel cells and high energy density batteries are appealing in this regard as they offer clean energy. They are attractive for a variety of power needs ranging from portable electronic devices to electric vehicles to stationary power.

Batteries are the major power sources for portable electronic devices. The exponential growth of popular portable electronic devices such as cellular phones and laptop computers has created an increasing demand for compact, lightweight power sources. Lithium-ion batteries have become appealing in this regard as they offer higher energy density compared to other rechargeable systems. The higher energy density also makes them attractive for hybrid and electric vehicles. Commercial lithium-ion cells currently use the layered lithium cobalt oxide as the cathode and carbon as the anode. However, the practical capacity of lithium cobalt oxide is limited, and cobalt is relatively expensive and toxic. On the other hand, the currently used graphite anode exhibits irreversible capacity loss. These difficulties have created enormous worldwide interest to develop alternative cathode and anode materials.

Unlike the battery technology, the fuel cell technology has not quite matured, and it is confronted with materials issues and high cost. For example, the high operating temperature of the conventional zirconia-based electrolyte limits the choice of electrode and interconnect materials for solid oxide fuel cells (SOFC). There is immense interest to lower the operating temperature of SOFC and use hydrocarbon fuels directly without requiring external reformers. The research activities focus on the development of compatible cathode, anode, and electrolyte combinations that can operate at an intermediate temperature of around 700°C. Additionally, cost-effective manufacturing and development of stable seals are critical for the success of the SOFC technology. In this regard, development of new materials as well as processing and characterization of ceramic materials play a key role.

To bring the ceramics community up to date on the fuel cell and battery technologies, the American Ceramic Society has been hosting symposia on related topics since 1995. This volume consists of 14 papers that were presented at the 106th Annual Meeting of the American Ceramic Society, Indianapolis, IN, April 18-21, 2004. A number of leading experts in materials science and engineering, solid state chemistry and physics, electrochemical science and technology from academia, industry and national laboratories presented their research and developments at this symposium. The presentations covered development of new materials and a fundamental understanding of the structure-property-performance relationships and the associated electrochemical and solid state phenomena. The symposium was sponsored by the Electronics, Basic Science, and Nuclear and Environmental divisions of the American Ceramic Society. Of the papers presented at this symposium, including several invited talks, 14 peer-reviewed papers are included in this volume under two subtopics: solid oxide fuel cells and lithium-ion batteries.

The editors acknowledge the support of several members of the Electronics, Basic Science, and Nuclear and Environmental divisions of the American Ceramic Society. The editors also thank all the authors, session chairs, manuscript reviewers, and the society staff who made the symposium and the proceeding volume a success. It is the sincere hope of the editors that the readers will appreciate and benefit from this collection of articles in the area of solid oxide fuel cells and lithium ion batteries.

Arumugam Manthiram
Prashant N. Kumta
S. K. Sundaram
Siu-Wai Chan

Solid Oxide Fuel Cells

CHARACTERIZATION OF Sr-DOPED NEODYMIUM COBALT OXIDE CATHODE MATERIALS FOR INTERMEDIATE TEMPERATURE SOLID OXIDE FUEL CELLS

K.T. Lee and A. Manthiram
Materials Science and Engineering Program
University of Texas at Austin
Austin, TX 78712

ABSTRACT

With an aim to explore as cathode materials for intermediate temperature solid oxide fuel cells, the structure and properties of $Nd_{1-x}Sr_xCoO_{3-\delta}$ oxides have been investigated for $0 \leq x \leq 0.5$. The $Nd_{1-x}Sr_xCoO_{3-\delta}$ system crystallizing in the orthorhombic perovskite structure exhibits a semiconductor to metal transition at $x \approx 0.3$, and the electrical conductivity increases with x. The thermal expansion coefficient decreases initially with increasing x, reaches a minimum at x = 0.3, and then increases. While the increasing electrical conductivity leads to an increase in electrocatalytic activity and power density initially with x, the enhanced interfacial reactions between the cathode and the $La_{0.8}Sr_{0.2}Ga_{0.8}Mg_{0.2}O_{2.8}$ (LSGM) electrolyte result in a decline in the activity at higher values of x > 0.4. Thus the x = 0.4 sample exhibits the highest catalytic activity with a maximum power density value of 0.24 W/cm^2 at 800 °C in single cells fabricated with the Ni-$Ce_{0.9}Gd_{0.1}O_{1.95}$ (GDC) cermet anodes.

INTRODUCTION

Solid oxide fuel cells (SOFCs) based on yttria-stabilized zirconia (YSZ) electrolytes generally need an operating temperature of around 1000 °C, which leads to difficulties arising from thermal expansion mismatch, interfacial reaction between the electrolyte and electrode, and limitations in the choice of interconnect materials.[1,2] These difficulties have generated interest in the development of SOFCs that can operate at an intermediate temperature of 500-800 °C, but the lower temperatures lead to large over-potential at the oxygen reduction electrode. Although the Sr-doped $LaMnO_3$ (LSM) perovskite oxides exhibit acceptable electrochemical activity at 1000 °C with YSZ, the low oxide ion conductivity prevent their use for intermediate temperature SOFCs.[3-5] With this respective, the Sr-doped $LaCoO_3$ (LSC) perovskite oxides have drawn attention due to their high electronic and ionic conductivities, but they experience high thermal expansion and low chemical stability.[6,7] The replacement of La by other lanthanides Ln = Sm and Gd can lower the thermal expansion due to the decrease in the ionicity of the Ln-O bond. However, replacement of the larger La by smaller Ln = Pr, Sm, and Gd will lower the electronic conductivity due to a bending of the O-Co-O bonds from 180° and a consequent decrease in the bandwidth.[8-11] To realize a compromise between these two parameters, we focus on the $Nd_{1-x}Sr_xCoO_{3-\delta}$ compositions since the ionic size of Nd^{3+} is intermediate between those of La^{3+} and Sm^{3+} or Gd^{3+} and the ionicity of the Nd-O bond is intermediate between those of La-O and Sm-O or Gd-O bonds. We present here the crystal chemistry, electrical conductivity, thermal expansion, and electrochemical performance of $Nd_{1-x}Sr_xCoO_{3-\delta}$ for $0 \leq x \leq 0.5$.

EXPERIMENTAL

The $Nd_{1-x}Sr_xCoO_{3-\delta}$ samples were synthesized by firing required amounts of Nd_2O_3, $SrCO_3$, and Co_3O_4 in air first at 900 °C for 12 h, followed by regrinding, pressing into pellets, and sintering at 1200 °C for 24 h. In order to study the effect of material synthesis procedure, the $Nd_{0.6}Sr_{0.4}CoO_{3-\delta}$ composition was also synthesized by a coprecipitation method. For the coprecipitation method, required amounts of Nd_2O_3, $SrCO_3$, and $Co(CH_3COO)_3{\cdot}4H_2O$ were dissolved in dilute nitric acid and the metal ions were then coprecipitated as carbonates and hydroxides by adding a mixture of KOH and K_2CO_3.[12] The coprecipitate was washed with deionized water, dried, fired at 500 °C for 5 h, ground, pressed into pellets, and sintered at 1200 °C for 24 h. The NiO-$Ce_{0.9}Gd_{0.1}O_{1.95}$ (GDC) cermet (Ni:GDC = 70:30 vol %) anode was synthesized by the glycine-nitrate combustion method.[13] Glycine was added to a nitric acid solution containing stoichiometric amounts of Gd_2O_3, CeO_2, and $Ni(CH_3COO)_2{\cdot}4H_2O$. The metal nitrate/glycine solution was heated on a hot plate to evaporate excess water and the anode cermet powder was obtained by self-sustaining combustion of the solid mass. The residual carbon was then removed by calcination at 600 °C for 3 h in air. The $La_{0.8}Sr_{0.2}Ga_{0.8}Mg_{0.2}O_{2.8}$ (LSGM) electrolyte was prepared by firing required amounts of La_2O_3, $SrCO_3$, Ga_2O_3, and MgO at 1100 °C for 5 h, followed by palletizing and sintering at 1500 °C for 10 h.

Crystal chemistry characterizations were carried out with X-ray diffraction (XRD) employing Rietveld method. Thermogravimetric analysis (TGA) and thermal expansion measurement were carried out with a heating/cooling rate of, respectively, 2 and 10 °C/min in air. The electrical conductivity data were collected with a four-probe dc method in the temperature range of 200-900 °C in air. Electrochemical performances were carried out with single cells at 800 °C. The $Nd_{1-x}Sr_xCoO_{3-\delta}$ cathodes and NiO-GDC cermet anode were prepared by screen printing on to a 1 mm thick LSGM electrolyte pellet, followed by firing for 3 h at 1000 °C for the cathode and 1200 °C for the anode. The geometrical area of the electrode was 0.25 cm^2 and Pt paste was used as the reference electrode. Humidified H_2 (~3% H_2O at 30 °C) and air were supplied as fuel and oxidant, respectively, at a rate of 100 cm^3/min.

RESULT AND DISCUSSION

All the $Nd_{1-x}Sr_xCoO_{3-\delta}$ ($0 \leq x \leq 0.5$) samples synthesized were found to be single phase and were indexed with the orthorhombic $GdFeO_3$ type perovskite structure (space group Pbnm, No. 62). The variations in the lattice parameters and lattice volume with Sr content x are shown in Figure 1. The lattice parameters and lattice volume increase with increasing x. The substitution of Sr^{2+} (ionic radius = 1.44 Å) for Nd^{3+} (1.27 Å) causes an oxidation of the larger Co^{3+} (0.61 Å) to a smaller Co^{4+} (0.53 Å) and/or the formation of oxygen vacancies. Nevertheless, a significantly larger size of Sr^{2+} compared to that of Nd^{3+} results in an overall increase in the lattice parameters and volume with x.

When Sr^{2+} ions are substituted for Nd^{3+} ions, the charge imbalance could be compensated by either or both of the following two mechanisms: electronic compensation by an oxidation of Co^{3+} to Co^{4+} in $Nd_{1-x}^{3+}Sr_x^{2+}Co_{1-x}^{3+}Co_x^{4+}O_3$ and/or ionic compensation by the formation of oxygen vacancies in $Nd_{1-x}^{3+}Sr_x^{2+}Co^{3+}O_{3-0.5x}$. Petrov et al.[6] reported that the concentration of Co^{4+} in the $La_{1-x}Sr_xCoO_{3-\delta}$ system increases with x, reaches a maximum at $x \sim 0.4$, and then decreases at higher values of x. This result suggests that at higher doping levels ($x > 0.4$), the charge imbalance is primarily compensated by the formation of oxygen vacancies. In order to verify the defect structure of $Nd_{1-x}Sr_xCoO_{3-\delta}$, we investigated the samples with thermogravimetric analysis (TGA) in air, and the TGA data of $Nd_{1-x}Sr_xCoO_{3-\delta}$ on heating in air are shown in Figure 2.

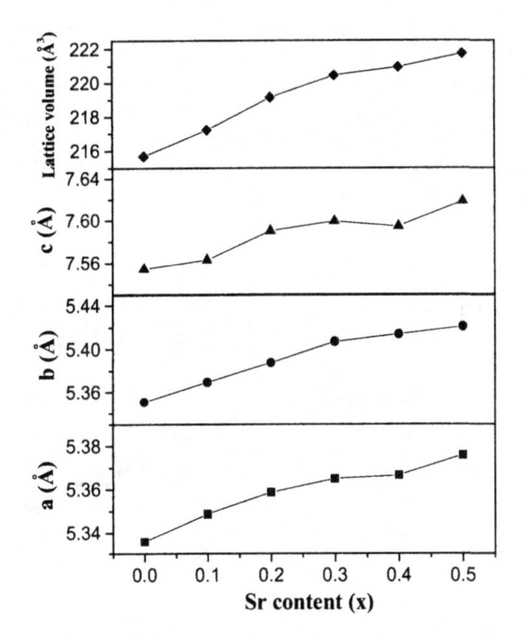

Figure 1. Variations of the lattice parameters and lattice volume of $Nd_{1-x}Sr_xCoO_{3-\delta}$ ($0 \leq x \leq 0.5$) with Sr content, x.

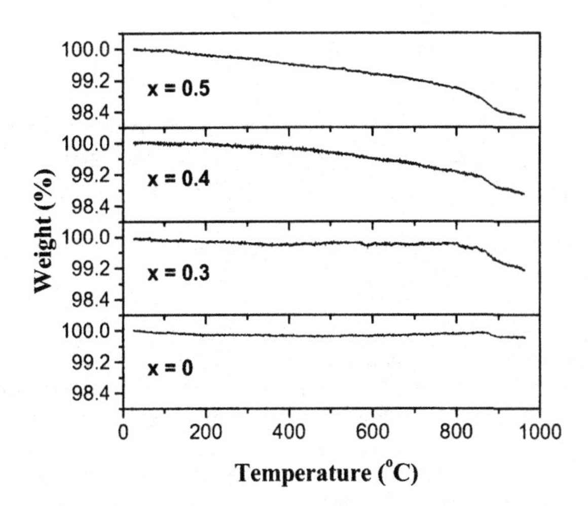

Figure 2. TGA plots of $Nd_{1-x}Sr_xCoO_{3-\delta}$ ($0 \leq x \leq 0.5$) recorded in air with a heating rate of 2 °C/min.

All the compositions show weight losses above 900 °C due to the thermal reduction of Co^{3+} or Co^{4+} to lower valence states and the consequent formation of oxide ion vacancies. While there is no detectable weight loss below 900 °C for x < 0.4, the x = 0.4 and 0.5 compositions show a large weight loss at relatively lower temperature (T > 300 °C), suggesting the ready formation of oxide ion vacancies in the latter. It appears that while electronic charge compensation mechanism may be dominant for x < 0.4, the ionic charge compensation mechanism may be dominant for x ≥ 0.4. In other words, the $Nd_{1-x}Sr_xCoO_{3-\delta}$ system could accommodate oxide ion vacancies and exhibit high oxide ion conductivity for x ≥ 0.4.

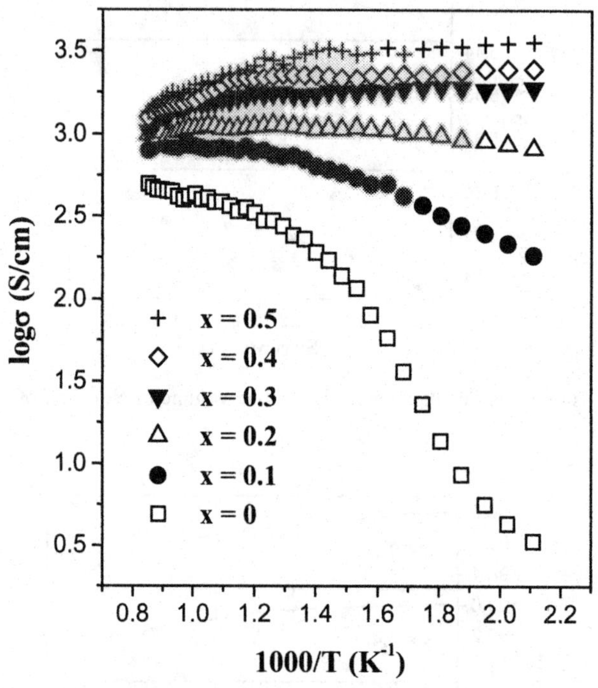

Figure 3. Variations of the electrical conductivity measured in air of $Nd_{1-x}Sr_xCoO_{3-\delta}$ ($0 \leq x \leq 0.5$) with temperature for various value of x.

The temperature dependence of the electrical conductivity of $Nd_{1-x}Sr_xCoO_{3-\delta}$ is shown in Figure 3. The conductivity increases with increasing Sr content, x, due to an increasing amount of the charge carriers. The faster decrease in conductivity at higher temperatures for samples with x ≥ 0.3 could be due to the loss of oxygen from the lattice at higher temperatures as indicated by the TGA data (Figure 2) and the consequent decrease in the Co^{4+} content and carrier concentration. The conductivity increases with increasing temperature for x < 0.3, implying a small polaron semiconductor behavior, but decreases with increasing temperature for x ≥ 0.3, implying a metallic behavior. The $Nd_{1-x}Sr_xCoO_{3-\delta}$ system thus exhibits a semiconductor to metal transition around x = 0.3. Torrance et al.[14] reported that the semiconductor-metal transition in

$ANiO_3$ (A = Pr, Nd, Sm, and Eu) perovskite oxides are related to the size of A-site cation. They also pointed out that as the size of the A-site cation increases, the tolerance factor increases and the O-Ni-O bond angle gradually straightens out, resulting in an increasing bandwidth and metallic conduction. Considering the $Nd_{1-x}Sr_xCoO_{3-\delta}$ system, the substitution of a larger Sr^{2+} for Nd^{3+} not only relieves the tensile stress in the (Nd,Sr)-O bond but also relieves the compressive stress in the Co-O bond due to an oxidation of the larger Co^{3+} to smaller Co^{4+} ions. This leads to an increase in the tolerance factor and the O-Co-O bond angle toward 180°, which manifests in an increase in the overlap between the Co-3d and O-2p orbitals and bandwidth, resulting in metallic conduction.

Table I. Total conductivity, average linear thermal expansion coefficient, BET surface area, and average crystallite size of $Nd_{1-x}Sr_xCoO_{3-\delta}$

x	Total conductivity at 800 °C (S/cm)	Average linear thermal expansion coefficient ($^{o}C^{-1}$)	BET surface area (m^2/g)	Average crystallite size* (Å)
0	449.1	27.8×10^{-6}	4.05	574
0.1	840.6	23.6×10^{-6}	5.02	594
0.2	980.7	19.7×10^{-6}	5.51	573
0.3	1427.4	17.8×10^{-6}	4.74	530
0.4	1500.7	18.3×10^{-6}	4.47	438
0.5	1769.8	18.9×10^{-6}	5.25	582

*Average crystallite size was estimated from the line broadening of X-ray diffraction peaks.

The total conductivity, average linear thermal expansion coefficient, BET surface area, and average crystallite size values are listed in Table I for $Nd_{1-x}Sr_xCoO_{3-\delta}$. TEC decreases with increasing x, reaches a minimum value of 17.8×10^{-16} $^{o}C^{-1}$ at x = 0.3, and then increases. Similar thermal expansion behaviors have been observed in the $Sm_{1-x}Sr_xCoO_{3-\delta}$ and $Pr_{1-x}Sr_xCoO_{3-\delta}$ systems.[9,11] Kostogloudis et al.[11] explained this behavior with Grüneisen's law,[15]

$$\alpha_V = \frac{\gamma C_V}{VB} \tag{1}$$

where α_V is the bulk TEC, γ is the Grüneisen's constant, C_v is the heat capacity at constant volume, V is the cell volume, and B is the bulk modulus. Since the cell volume increases with Sr substitution, it causes a decrease in TEC with x. However, the changes in valence and spin states of cobalt ions could also be useful parameters in understanding the variations in TEC. It has been reported[16-18] that Co^{3+} ions in $LnCoO_3$ (Ln: La, Nd, and Gd) are all in the diamagnetic low spin Co^{III} state ($t_{2g}^6 e_g^0$, ionic radius r = 0.545 Å) at low temperature, and transform to a paramagnetic high spin Co^{3+} state ($t_{2g}^4 e_g^2$, r = 0.61 Å) or an intermediate spin Co^{iii} state ($t_{2g}^5 e_g^1$) at higher temperatures. Considering the ionic radii values, these spin transitions can lead to a larger thermal expansion coefficient for $NdCoO_3$. However, the substitution of Sr^{2+} for Nd^{3+} oxidizes Co^{3+} to Co^{4+}, which exists in the low spin Co^{IV} state ($t_{2g}^5 e_g^0$, r = 0.53 Å).[18] Accordingly, the TEC of $Nd_{1-x}Sr_xCoO_{3-\delta}$ decreases with increasing x for $0 \le x \le 0.3$ due to a decrease in the Co^{3+} concentration and the consequent decrease in the amount of low spin to high spin transition.

However, at higher doping levels with $x \geq 0.4$, Co^{4+} is reduced to Co^{3+} and oxygen vacancies are formed by the ionic charge compensation mechanism. The formation of oxygen vacancies could cause a decrease in the electrostatic attractive forces[19,20] and thereby an increase in TEC.

The BET surface area and the average crystallite size of the $Nd_{1-x}Sr_xCoO_{3-\delta}$ powders do not vary significantly with x (Table I). Therefore, changes in geometrical morphologies with x may not have a significant influence on the properties in the $Nd_{1-x}Sr_xCoO_{3-\delta}$ system investigated here. Figure 4 shows the variations of the power density and the over-potential with current density for the various $Nd_{1-x}Sr_xCoO_{3-\delta}$ cathode compositions at 800°C. The power density increases with increasing Sr content for $0 \leq x \leq 0.4$, reaches a maximum at $x = 0.4$, and then decreases for $x = 0.5$. The over-potential decreases with x initially, reaches a minimum at $x = 0.4$, and then increases similar to the trend in the power density. The significant decrease of the power density and increase of over-potential for $x = 0$ and 0.1 is attributed to the lower electrical conductivity as well as surface cracks due to the large thermal expansion coefficient. The main reaction at cathode is the oxygen reduction reaction given below:

$$\frac{1}{2}O_2\,(gas) + 2e^-\,(cathode) \rightarrow O^{2-}\,(electrolyte) \tag{2}$$

It is usually assumed that there are two possible paths for the cathode reaction. One is the surface path and the other is the bulk path. In surface path, oxygen is adsorbed and partly reduced on the surface of the cathode, and diffuses along the cathode surface to the three phase boundary (TPB, cathode/electrolyte/gas) where it becomes fully reduced and completes the ionic transfer into the electrolyte. This surface path for the oxide ion incorporation into the electrolyte is usually considered as a favorable mechanism. Under electrical load, the oxide ion could also diffuse through the cathode and be incorporated into the electrolyte at the cathode/electrolyte interface (bulk path). Considering the defect structure described earlier, the bulk path is unfavorable for the $x = 0$, 0.1, 0.2, and 0.3 samples due to the rather low oxygen deficiency, resulting in lower catalytic activity. On the other hand, the $x = 0.4$ and 0.5 samples with significant amount of oxide ion vacancies and mixed electronic-ionic conduction behaviors are expected to show good electrochemical performance due to bulk diffusion. However, there is a decrease in the electrochemical performance on going from $x = 0.4$ to $x = 0.5$ due to an enhanced interfacial reaction between the LSGM electrolyte and the $Nd_{0.5}Sr_{0.5}CoO_{3-\delta}$ cathode. As a result, the $x = 0.4$ sample shows the maximum power density with the lowest over-potential value in the $Nd_{1-x}Sr_xCoO_{3-\delta}$ system.

The interfacial reaction for the $x = 0.5$ sample is evident on examining the X-ray powder diffraction patterns recorded after heating the $Nd_{1-x}Sr_xCoO_{3-\delta}$ and LSGM mixtures at 1000 °C for 3 h (Figure 5). While no reaction product is observed for the compositions with $x \leq 0.4$, the insulating $LaSrGa_3O_7$ reaction product is found for the $x = 0.5$ composition. Huang et al.[21] also reported the formation of $LaSrGa_3O_7$ as a result of Co diffusion at the interface between $La_{0.5}Sr_{0.5}CoO_{3-\delta}$ and LSGM after heating at 1050 °C for 2 h. The diffusion of Co into the LSGM lattice results in a destabilization of its ability to hold high Sr concentration and consequently in the formation of the reaction product $LaSrGa_3O_7$.

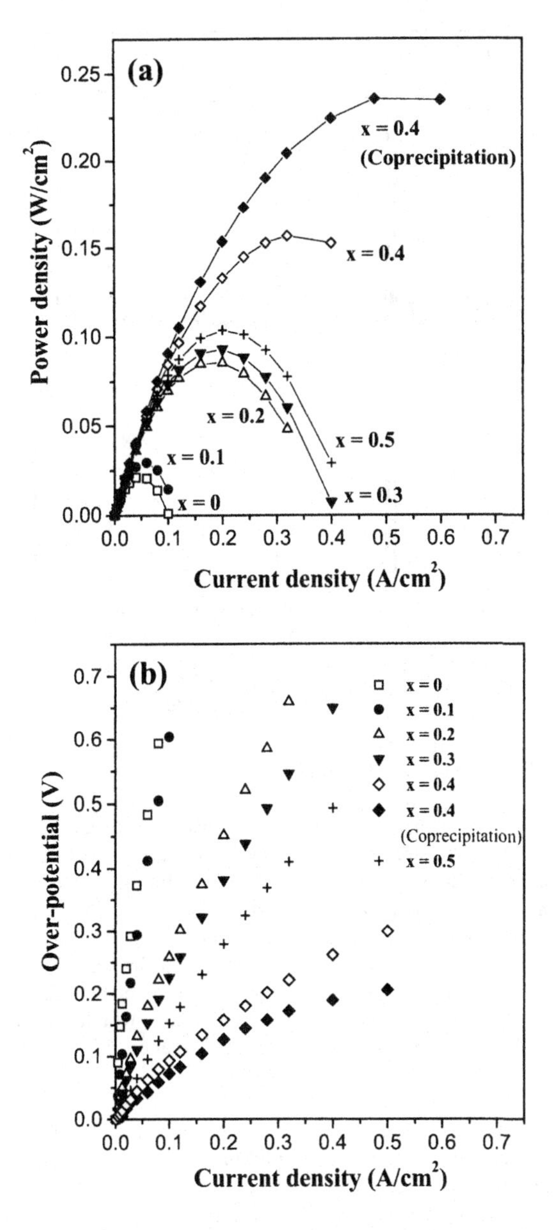

Figure 4. Electrochemical performance data of $Nd_{1-x}Sr_xCoO_3$/LSGM/Ni-GDC single cells at 800 °C: Variations of (a) power density and (b) cathode over-potential.

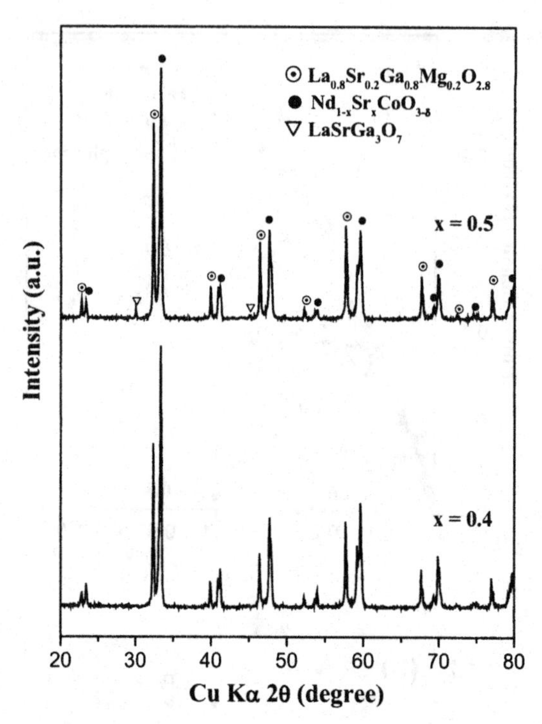

Figure 5. X-ray powder diffraction patterns recorded after heating the $Nd_{1-x}Sr_xCoO_{3-\delta}$ (x = 0.4 and 0.5) cathode and the LSGM electrolyte powders at 1000 °C for 3 h.

In order to study the effect of material synthesis procedure on the electrochemical performance, we have also compared the performances of the x = 0.4 compositions that were prepared by a standard solid state reaction and a coprecipitation method. As shown in Figure 4, the single cell fabricated with the coprecipitated $Nd_{0.6}Sr_{0.4}CoO_{3-\delta}$ cathode exhibits a lower over-potential and a higher maximum power density value of 0.24 W/cm^2 at 800 °C than that prepared by the solid state reaction. While the solid state reacted cathode has only point contacts with each particle and poor adhesion between the cathode and electrolyte, the cathode obtained by the coprecipitation method has area contacts and good adhesion, which lead to low contact and charge transfer resistances and good electrochemical performance.

CONCLUSIONS

$Nd_{1-x}Sr_xCoO_{3-\delta}$ perovskite oxides have been synthesized for $0 \leq x \leq 0.5$ and characterized as cathode materials for intermediate temperature solid oxide fuel cells. All the compositions have an orthorhombic structure, and the lattice volume and tolerance factor increase with increasing Sr content. While an electronic charge compensation mechanism is dominant for $x \leq 0.3$, the x = 0.4

and 0.5 compositions have oxygen deficient structures resulting in mixed electronic-ionic conduction behavior. The spin states of the cobalt ions and the internal stresses resulting from a bond length mismatch between the (Nd,Sr)-O and Co-O bonds play an important role in the structure and thermal expansion behaviors. The electrical conductivity increases with x due to an increasing amount of charge carriers and a semiconductor to metal transition occurs at $x \approx 0.3$. The semiconductor to metal transition originates due to an increase in the O-Co-O bond angle towards 180° and a consequent increase in the bandwidth with Sr doping. The $x = 0.4$ composition with the mixed conducting behavior shows the best electrochemical performance. Although the $x = 0.5$ sample exhibits both high electronic and oxide ion conductivities, the interfacial reaction with the LSGM electrolyte resulting in the formation of the unwanted $LaSrGa_3O_7$ leads to a decline in the electrochemical performance compared to the $x = 0.4$ sample. The $x = 0.4$ sample obtained by a coprecipitation method shows the maximum power density value of 0.24 W/cm^2 at 800 °C in single cells fabricated with 1 mm LSGM electrolyte, Ni-$Ce_{0.9}Gd_{0.1}O_{1.95}$ (GDC) cermet anode, and $Nd_{0.6}Sr_{0.4}CoO_{3-\delta}$ cathode.

ACKNOWLEDGEMENT

This work was supported by the Office of Naval Research through the Electric Ship Research and Development Consortium and Welch Foundation Grant F-1254.

REFERENCES

[1]N.Q. Minh, "Ceramic Fuel Cells," *Journal of the American Ceramic Society*, **76** [3] 563-88 (1993).

[2]T. Tsai and S.A. Barnett, "Effect of LSM-YSZ Cathode on Thin-Electrolyte Solid Oxide Fuel Cell Performance," *Solid State Ionics*, **93** [3-4] 207-17 (1997).

[3]T. Horita, K. Yamaji, M. Ishikawa, N. Sakai, H. Yokokawa, T. Kawada, and T. Kato, "Active Sites Imaging for Oxygen Reduction at the $La_{0.9}Sr_{0.1}MnO_{3-x}$/Yttria-Sabilized Zrconia Interface by Secondary-Ion Mass Spectrometry," *Journal of the Electrochemical Society*, **145** [9] 3196-202 (1998).

[4]F. Zheng and L.R. Pederson, "Phase Behavior of Lanthanum Manganites," *Journal of the Electrochemical Society*, **146** [8] 2810-16 (1999).

[5]R.A. De Souza and J.A. Kilner, "Oxygen Transport in $La_{1-x}Sr_xMn_{1-y}Co_yO_{3\pm\delta}$ in Perovskites: Part I. Oxygen Tracer Diffusion," *Solid State Ionics*, **106** [3-4] 175-87 (1998).

[6]A.N. Petrov, O.F. Kononchuk, A.V. Andreev, V.A. Cherepano, and P. Kofstad, "Crystal Structure, Electrical and Magnetic Properties of $La_{1-x}Sr_xCoO_{3-y}$," *Solid State Ionics*, **80** [3-4] 189-99 (1995).

[7]S.B. Adler, "Mechanism and Kinetics of Oxygen Reduction on Porous $La_{1-x}Sr_xCoO_{3-\delta}$ Electrodes," *Solid State Ionics*, **111** [1-2] 125-34 (1998).

[8]Y. Takeda, H. Ueno, N. Imanishi, O. Yamamoto, N. Sammes, and M.B. Phillipps, "$Gd_{1-x}Sr_xCoO_3$ for the Electrode of Solid Oxide Fuel Cells," *Solid State Ionics*, **86-88** [2] 1187-90 (1996).

[9]H.Y. Tu, Y. Takeda, N. Imanishi, and O. Yamamoto, "$Ln_{1-x}Sr_xCoO_3$ (Ln = Sm, Dy) for the Electrode of Solid Oxide Fuel Cells," *Solid State Ionics*, **100** [3-4] 283-88 (1997).

[10]M. Koyama, C. Wen, T. Masuyama, J. Otomo, H. Fukunaga, K. Yamade, K. Eguchi, and H. Takahashi, "The Mechanism of Porous $Sm_{0.5}Sr_{0.5}CoO_3$ Cathodes Used in Solid Oxide Fuel Cells," *Journal of the Electrochemical Society*, **148** [7] A795-A801 (2001).

[11]G.Ch. Kostogloudis, N. Vasilakos, and Ch. Ftikos, "Crystal Structure, Thermal and Electrical Properties of $Pr_{1-x}Sr_xCoO_{3-\delta}$ (x = 0, 0.15, 0.3, 0.4, 0.5) Perovskite Oxides," *Solid State Ionics*, **106** [3-4] 207-18 (1998).

[12]J.P. Tang, R.I. Dass, and A. Manthiram, "Comparison of the Crystal Chemistry and Electrical Properties of $La_{2-x}A_xNiO_4$ (A = Ca, Sr, and Ba)," *Materials Research Bulletin*, **35** 411-24 (2000).

[13]L.A. Chick, L.R. Pedersen, G.D. Maupin, J.L. Bates, L.E. Thomas, and G.J. Exarhos, "Glycine-Nitrate Combustion Synthesis of Oxide Ceramic Powders," *Materials Letters*, **10** [1-2] 6-12 (1990).

[14]Torrance, P. Lacorre, A. I. Nazzal, E. J. Ansaldo, and Ch. Niedermayer, "Systematic Study of Insulator-Metal Transitions in Perovskites $RNiO_3$ (R = Pr, Nd, Sm, Eu) Due to Closing of Charge-transfer Gap," *Physical Review B*, **45** [14] 8209-12 (1992)

[15]J.N. Eastabrook, "Thermal Expansion of Solids at High Temperatures," *The Philosophical Magazine*, **2** [24] 1421-26 (1957).

[16]P.M. Raccah and J.B. Goodenough, "First-Order Localized-Electron ⇆ Collective-Electron Transition in $LaCoO_3$," *Physical Review*, **155** [3] 932-43 (1967).

[17]D.S. Rajoria, V.G. Bhide, G. Rama Rao, and C.N. R.Rao, "Spin State Equilibria and Localized versus Collective d-Electron Behaviour in Neodymium and Gadolinium Trioxocobaltate (III)," *Journal of the Chemical Society Faraday Transactions II*, **70** [3] 512-23 (1974).

[18]M.A. Señaris-Rodriguez and J.B. Goodenough, "Magnetic and Transport Properties of the System $La_{1-x}Sr_xCoO_{3-\delta}$ ($0 < x \leq 0.50$)," *Journal of Solid State Chemistry*, **118** 323-36 (1995).

[19]H. Hayashi, M. Suzuki, and H. Inaba, "Thermal Expansion on Sr- and Mg-Doped $LaGaO_3$," *Solid State Ionics*, **128** [1-4] 131-39 (2000).

[20]H. Ullmann, N. Trofimenko, F. Tietz, D. Stöver, and A. Ahmad-Khanlou, "Correlation between Thermal Expansion and Oxide Ion Transport in Mixed Conducting Perovskite-Type Oxides for SOFC Cathodes," *Solid State Ionics*, **138** [1-2] 79-90 (2000).

[21]K. Huang, M. Feng, and J.B. Goodenough and M. Schmerling, "Characterization of Sr-Doped $LaMnO_3$ and $LaCoO_3$ as Cathode Materials for a Doped $LaGaO_3$ Ceramic Fuel Cell," *Journal of the Electrochemical Society*, **143** [11] 3630-35 (1996).

MICROSTRUCTURE AND ELECTRICAL CONDUCTIVITY STUDIES OF $(La,Sr)(Cr,Mn,Co)O_3$

Soydan Ozcan and Rasit Koc
Department of Mechanical Engineering and Energy Processes
Southern Illinois University
Carbondale, IL, 62901-6603

ABSTRACT

The $LaCrO_3$-$LaMnO_3$-$LaCoO_3$ system was investigated in order to develop an improved cathode material for solid oxide fuel cells (SOFCs). The formation of a solid solution by the ternary oxide mixture was confirmed by the X-ray diffraction pattern. Sintering studies conducted in air at 1500 °C produced compositions at the center of the ternary which exhibited 95% of theoretical density. The AC electrical conductivity of compositions measured in air from 373K to 1073K increased significantly as Co content increased. At 1073K, the electrical conductivity ranged from 13 S/cm for $La_{0.5}Sr_{0.5}Cr_{0.7}Mn_{0.1}Co_{0.2}O_3$ to 55 S/cm for $La_{0.9}Sr_{0.1}Cr_{0.5}Mn_{0.1}Co_{0.4}O_3$. The results observed during sintering and AC electrical conductivity testing provide a basis for evaluation of the mixed material for practical applications.

INTRODUCTION

Perovskite-type oxides have been successfully used in various applications including high temperature heating elements, sensor materials, electrodes, solid oxide fuel cells (SOFCs) and others. The need to develop clean power generators and improve electricity production has revived the significance of SOFCs. The energy produced in SOFCs is a result of the direct conversion of chemical energy to electrical energy via an electrochemical reaction. SOFCs are an economical, reliable, and clean electrochemical generator.

Perovskite-type oxides have the general formula (ABO_3). In this formula the large cation site A may be an alkali, alkaline earth, or rare earth ion while B represents a transition metal for cation. ABO_3-based compounds usually deviate from the ideal cubic perovskite structure and form orthorhombic, rhombohedral, or tetragonal structures [1]. Electrical transport properties of substitutionally mixed perovskites-type oxides have been extensively studied. Recently studied examples include $La(Cr,Mn)O_3$ and $La(Cr,Co)O_3$ compositions. These investigations have indicated that end members of both compositions exhibit a thermally activated high-temperature electrical conductivity due to small-polaron hopping among the B site cations [1,2]. Electrical conductivity is expressed as

$$\sigma=(NC)e\mu \quad (1)$$

where N is the density of available sites, C is the fraction of available sites occupied by charge carriers, e is the electron charge, and μ is the mobility of charge carriers. Electrical conduction of p-type perovskite-type oxide may be increased by various methods: (i) Substituting donors or acceptors for the A lattice sites in the perovskite structure by controlling the small-polaron concentration found on the B lattice site cations; as described by the Verwey principle [3,4]. (ii) Increasing the carrier concentration by substituting different transition metals for the B lattice site [5]. (iii) Removing the oxygen ions to reduce the valence state of the remaining ions and subsequently reduce their coordination number [6].

In the present study, the temperature and compositional dependence of AC electrical conductivity exhibited by $(La,Sr)(Cr,Mn,Co)O_3$ perovskite compounds were examined. The experimental results of the electrical conductivity test were analyzed using the small-polaron model. The specimens prepared for use in this study were sintered in air.

EXPERIMENTAL PROCEDURE

Specimen used in the $(La,Ca)(Cr,Mn,Co)O_3$ ceramics of this study were prepared using a polymer precursor method similar to that first described by Pechini [7]. Perovskite-type compounds were synthesized from carbonates of lanthanum, strontium, and manganese; and nitrates of cobalt and chromium. The starting chemicals were reagent-grade materials. Thermogravimetric methods (TG) were used to determine the actual cation contents of the starting chemicals. The desired compositions were prepared by dissolving measured amounts of carbonates and nitrates in solutions of citric acid, ethylene glycol and distilled water. The mixtures were heated on a hot plate at approximately 95°C until polymerization had occurred. Subsequent heating at higher temperatures resulted in the composition of the polymer resin and allowed conversion into the required oxide. Final calcinations were conducted at 850°C. The resulting powders were milled and subjected to X-ray diffraction to ensure the presence of a single solid phase. For the electrical conductivity and sintering studies, the powders were pressed into pellets with the aid of PVA (Poly-vinyl alcohol) and water binder. Scanning electron micrographs of the polished and thermally etched surfaces of sintered specimen were obtained using scanning electron microscopy (SEM) performed utilizing Hitachi S570 (Tokyo, Japan). The materials were characterized at room temperature by X-Ray diffraction analysis (XRD) (Rigaku, Tokyo, Japan) with CuK_α radiation.

The specimens were cut in rectangular shape (0.3 x 0.3 x 2,0 cm) from the sintered bars and a thin film of platinum was applied as contacts on both side of the pellets to neutralize the effect of barrier layers and non-ohmic contacts. AC electrical conductivity measurements were performed in a high temperature tube furnace. The specimens were mounted between two platinum blocks inside an Al_2O_3 tube, which had type-S thermocouples (Pt-10% Rh/Pt) as electrical contacts. Electrical conductivity of the specimens was measured by AC Resistance Bridge (LR 700, Linear Research Inc., San Diego, CA, USA) using four-wire alternative current (AC) Lock-Balance™ technique. This method of measuring electrical conductivity eliminates the thermal electromotive force (EMF) DC voltage errors and also permits voltage resolution down to the sub-nanovolt region.

RESULTS AND DISCUSSIONS

The XRD patterns of sintered specimens are shown in Figure 1. These XRD results showed that there was no additional phase in the specimens, indicating that $LaCrO_3$-$LaMnO_3$-$LaCoO_3$ formed solid solution.

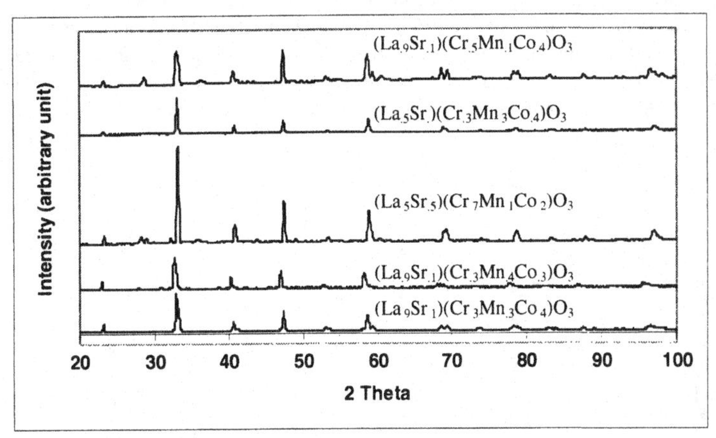

Fig. 1 XRD patterns of sintered specimens

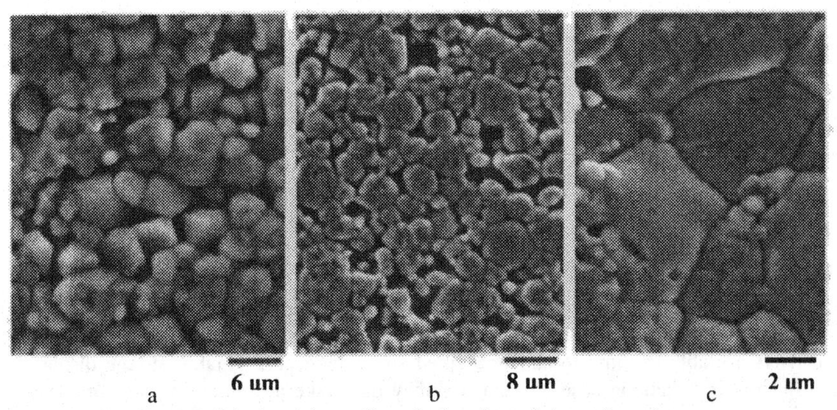

Fig. 2 Polished and thermally etched surface of sintered specimen,
a- $(La_{.9}Sr_{.1})(Cr_{.3}Mn_{.3}Co_{.4})O_3$ b- $(La_{.5}Sr_{.5})(Cr_{.7}Mn_{.1}Co_{.2})O_3$ c- $(La_{.9}Sr_{.1})(Cr_{.3}Mn_{.4}Co_{.3})O_3$

Scanning electron micrographs of the polished and thermally etched surfaces of the specimens are presented in Figure 2. The specimens displayed in the micrograph exhibit uniform grain shape. Additionally, grain-boundary morphology similar to a liquid phase may also be observed (rounding of grain corners). Detailed studies of this behavior in $LaCrO_3$-$LaCoO_3$ systems have been conducted in the past [8,9]. The nature of the liquid phase and sintering characteristics of the solid solution in the system will be reported in the upcoming publications.

AC electrical conductivity measurements of $(La,Sr)(Cr,Mn,Co)O_3$ compositions have been performed from 373K to 1073K in air. The temperature dependence of AC electrical conductivity of five compounds is displayed in Figure 3. The electrical conduction of the compounds is thermally activated over a wide range of temperatures. This behavior is expected and occurs as a result of small-polaron transport. Sr^{+2} ions are distributed randomly in the A lattice site in the strontium doped $La(Cr,Mn,Co)O_3$ composition. The presence at Sr^{+2} ions changes the small polaron concentration found on the B lattice site cations. A small-polaron is associated with the particular Cr, Mn, or Co site if the ion is in the +4 valance, rather than the stoichiometric +3 valance state. Charge transportation takes place via the B lattice site ions which are in the +4 valance states.

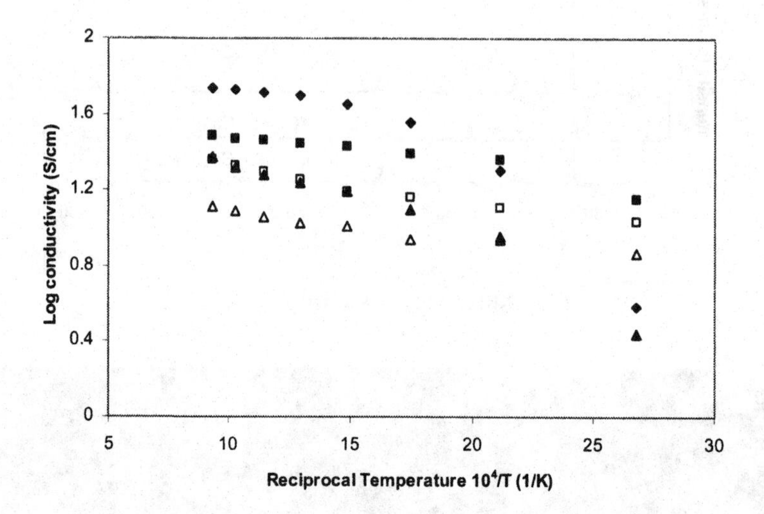

Fig.3 Electrical conductivity versus reciprocal temperature in air.
■:$(La_{.9}Sr_{.1})(Cr_{.3}Mn_{.3}Co_{.4})O_3$ ♦:$(La_{.9}Sr_{.1})(Cr_{.5}Mn_{.1}Co_{.4})O_3$ ▲:$(La_{.9}Sr_{.1})(Cr_{.3}Mn_{.4}Co_{.3})O_3$
Δ:$(La_{.5}Sr_{.5})(Cr_{.7}Mn_{.1}Co_{.2})O_3$ □: $(La_{.5}Sr_{.5})(Cr_{.3}Mn_{.3}Co_{.4})O_3$

Experimental results of the electrical conductivity measurements were analyzed using the small-polaron model. Two different type of small polaron have been determined. In the adiabatic case, carriers are able to jump to an unoccupied site whenever a suitable atomic displacement occurs. In the non-adiabatic case, the carriers may only take advantage of a favorable atomic configuration to facilitate a site change [10]. The temperature-dependent electrical conductivity of an adiabatic small-polaron material is expressed as

$$\sigma = \left[\frac{NC(1-C)e^2 va^2}{kT}\right]\exp\left(\frac{-E_A}{kT}\right) \tag{2}$$

where N is the density of available sites, C is the fraction of sites occupied by an electron or polaron, e is the unit charge, v is the optical phonon frequency, a is the lattice spacing, k is the Boltzman constant, T is absolute temperature, and E_A is the activation energy. The temperature dependent electrical conductivity of non-adiabatic material is expressed as

$$\sigma = \left[\frac{NC(1-C)e^2a^2}{kT}\right]\left(\frac{4\pi^2 j^2}{h^2}\right)\left(\frac{\pi}{4kTE_A}\right)^{1/2} \exp\left(\frac{-E_A}{kT}\right) \quad (3)$$

where j is the electron transfer energy, and h is the Planck's constant. If all the variables in the pre-exponential term of equations (2) and (3) are independent of temperature or slightly temperature-dependent a plot of log (σT) (for the adiabatic case) or log ($\sigma T^{3/2}$) (for the non-adiabatic case) versus reciprocal temperature should give a straight line with a gradient proportional to the activation energy associated with small-polaron hops.

Figures 4 and 5 shows the logarithmic plot of σT versus $10^4/T$ and $\sigma T^{3/2}$ versus $10^4/T$ for the (La,Sr)(Cr,Mn,Co)O_3 compositions covering the entire temperature range studied, respectively. Linear regression was used to calculate the activation energies for both adiabatic and non-adiabatic regimes and R^2 values were calculated to evaluate the linearity of plots for the compositions. The greater the value of R^2, the better the fit of the data to linear function. Arrhenius plots created for the Sr doped compositions were nearly linear over a wide range, as shown in Figures 4 and 5.

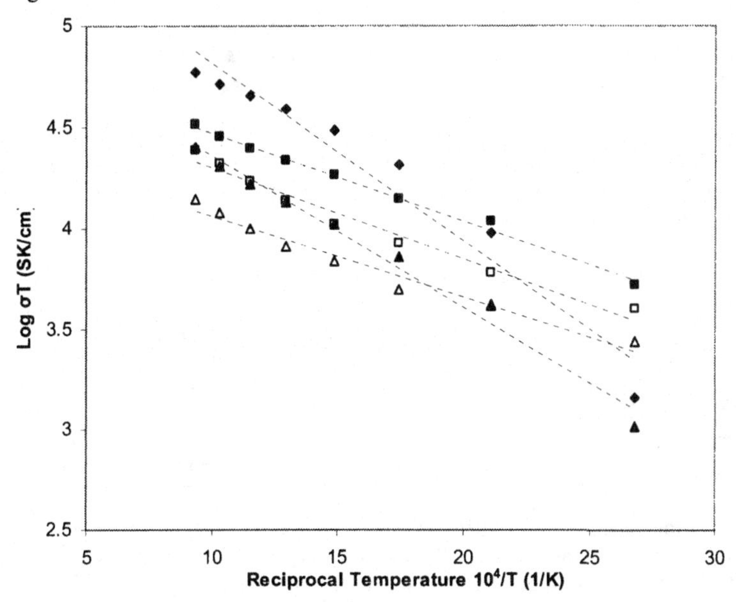

Fig 4. Log σT versus reciprocal temperature,
■:($La_{.9}Sr_{.1}$)($Cr_{.3}Mn_{.3}Co_{.4}$)O_3 ♦:($La_{.9}Sr_{.1}$)($Cr_{.5}Mn_{.1}Co_{.4}$)O_3 ▲:($La_{.9}Sr_{.1}$)($Cr_{.3}Mn_{.4}Co_{.3}$)O_3
Δ:($La_{.5}Sr_{.5}$)($Cr_{.7}Mn_{.1}Co_{.2}$)O_3 □: ($La_{.5}Sr_{.5}$)($Cr_{.4}Mn_{.3}Co_{.3}$)O_3

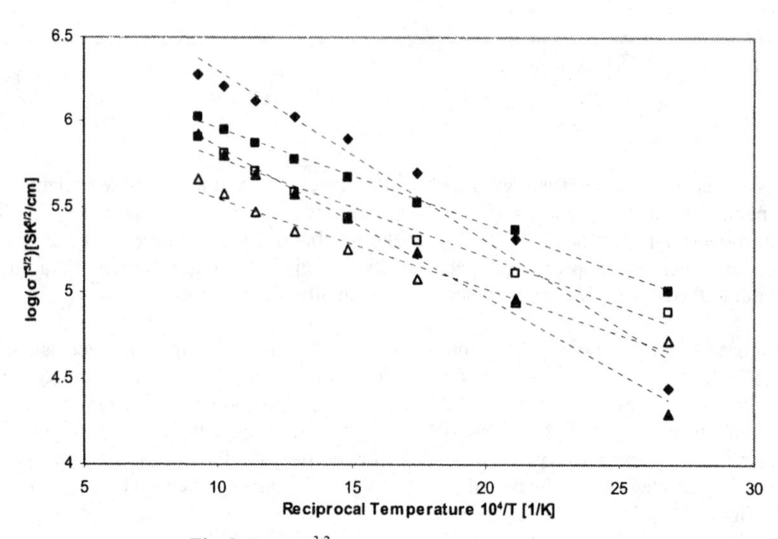

Fig 5. Log $\sigma T^{3/2}$ versus reciprocal temperature,
■:$(La_{.9}Sr_{.1})(Cr_{.3}Mn_{.3}Co_{.4})O_3$ ♦:$(La_{.9}Sr_{.1})(Cr_{.5}Mn_{.1}Co_{.4})O_3$ ▲:$(La_{.9}Sr_{.1})(Cr_{.3}Mn_{.4}Co_{.3})O_3$
Δ:$(La_{.5}Sr_{.5})(Cr_{.7}Mn_{.1}Co_{.2})O_3$ □: $(La_{.5}Sr_{.5})(Cr_{.4}Mn_{.3}Co_{.3})O_3$

The calculated activation energies and R^2 values for both adiabatic and non-adiabatic cases are given in Table 1. For all studied compositions, the high linearity of the plots is consistent with the assumption that electrical conduction in these compositions occurs via the non-adiabatic small polaron mechanism (see Table 1).

Table1 Activation energies for $(La,Sr)(Cr,Mn,Co)O_3$ compositions

Composition	Adiabatic case (meV)	Non-adiabatic case (meV)	Non-adiabatic case (R^2)	Non-adiabatic case (R^2)
$La_{0.9}Sr_{0.1}Cr_{0.3}Mn_{0.3}Co_{0.4}O_3$	43	56	0.9942	0.9961
$La_{0.9}Sr_{0.1}Cr_{0.3}Mn_{0.4}Co_{0.3}O_3$	75	88	0.9835	0.9909
$La_{0.9}Sr_{0.1}Cr_{0.5}Mn_{0.1}Co_{0.4}O_3$	87	101	0.9484	0.9671
$La_{0.5}Sr_{0.5}Cr_{0.4}Mn_{0.3}Co_{0.3}O_3$	45	58	0.9721	0.9734
$La_{0.5}Sr_{0.5}Cr_{0.7}Mn_{0.1}Co_{0.2}O_3$	40	53	0.9704	0.9750

The activation energies of studied $(La,Sr)(Cr,Mn,Co)O_3$ compositions are 56 meV for $La_{0.9}Sr_{0.1}Cr_{0.3}Mn_{0.3}Co_{0.4}O_3$ and 88 meV for $La_{0.9}Sr_{0.1}Cr_{0.3}Mn_{0.4}Co_{0.3}O_3$. Substitution of Sr for La in $La(Cr,Mn,Co)O_3$ composition should result in the formation of Cr^{+4}, Mn^{+4} and Co^{+4}. The formation of ions that are in the +4 valance state increases the small polaron concentration and decrease the activation energy for conduction.

Although the end members of the $(La,Sr)(Cr,Mn,Co)O_3$ system, ($LaCrO_3$, $LaMnO_3$, and $LaCoO_3$) have similar activation energies, they posses different electrical conductivity properties. $LaCoO_3$ exhibited the greatest electrical conductivity, followed by $LaMnO_3$ and $LaCrO_3$ [11,12].

Substitution of 10 mol% Mn for Co in $La_{0.9}Sr_{0.1}Cr_{0.3}Mn_{0.3}Co_{0.4}O_3$ resulted in a decrase in electrical conductivity and an increase in activation energy of approximately 32 meV. The higher Co concentration increases the electrical conductivity of the system due to the connection path available from Co sites.

CONCLUSION

Sintering studies in air showed that the compositions in the ternary sintered at 1500°C, 95% of theoretical density was achieved for the compositions. High temperature AC electrical conductivity measurements were performed on $(La,Sr)(Cr,Mn,Co)O_3$ system. The electrical conductivity measured in air from 373K to 1073K and the measured values at 1073K ranged from 13 S/cm for $La_{0.5}Sr_{0.5}Cr_{0.7}Mn_{0.1}Co_{0.2}O_3$ to 55S/cm for $La_{0.9}Sr_{0.1}Cr_{0.5}Mn_{0.1}Co_{0.4}O_3$. The results displayed thermally activated temperature dependence behavior between temperature 373K and 1073K, as described by small polaron hopping theory. Activation energies were calculated from Arrhenius plots and ranged from 53 meV to 101 meV. The evaluation of results indicated that electrical conduction of compositions occurs via the non-adiabatic small polaron mechanism.

ACKNOWLEDGEMENT

This research supported by Materials Technology Center at Southern Illinois University. Zachary Crothers is also acknowledged for his help in the review of the manuscript.

REFERENCES

[1]R. Raffaelle, H.U. Anderson, D.M. Sparlin and P.E. Parris, "Transport anomalics in the high temperature hopping conductivity and thermopower of Sr-doped $La(Cr,Mn)O_3$," *Physical Review B,* **43** 7991-99 (1991).

[2]H.Taguchi, M.Shimada and M.Koizumi, "Electrical properties in the system $(La_{1-x}Ca_x)CoO_3$ $(0.1 \leq x \leq 0.5)$", *Journal of Solid State Chemistry,* **44** [2] 254-56 (1982).

[3]E.J.Verwey, P.W.Haaijam, F.C.Romeijh and G.W. Van, "Controlled-Valency semiconductors" *Philips Research Reports,* **5** 173-87 (1950).

[4]S.R. Sehlin, H.U. Anderson, and D.M. Sparlin, "Semiempirical model for the electrical properties of $La_{1-x}Ca_xCo0_3$", *Physical Review B,* **52** 11681-89, (1995).

[5]R. Koc and H.U. Anderson, "Investigation of Sr-Doped La(Cr,Mn)O3 for solid oxide fuel cells", *Journal of Materials Science,* **27** 5837-43 (1992).

[6]I.O. Troyanchuk, S.V. Trukhanov, H. Szymczak, J.Przewoznick, and K.Barner, "Phase transition in $La_{1-x}Ca_xMnO_{3-x/2}$ manganites," *Journal of Experimental and Theoretical Physics,* **93** [1] 161-67 (2001).

[7]M. Pechini, "Method of preparing lead and alkaline earth titanates and niobates and coatings using the same to form a capacitor," U.S. Pat. No. 3 330 697 (1967).

[8]R. Koc and H.U. Anderson, "Liquid phase sintering of $LaCrO_3$", J. of the European Ceramic Society, **9** 285-292 (1992).

[9]R. Koc and H. Kaga "Investigation of sintering behavior of (La,Ca)(Cr,Co) O_3 using high temperature DSC", *Fifty European Solid Oxide Fuel Cell Forum Procedings,* **1** 409-16 (2002)

[10]R. Koc, H.U. Anderson, "Electrical and thermal properties of $(La,Ca)(Cr,Co)O_3$, *Journal of European Ceramic Society,* **15** 867-74 (1995).

[11]R. Koc and H.U.Anderson, "Electrical conductivity and seebeck coefficient of $(La,Ca)(Cr,Co)O_3$", *Journal of Material Science*, **27** 5477-82 (1992).

[12] R. Raffaelle, H.U. Anderson, D.M. Sparlin and P.E. Parris, "Evidence for a crossover from multiple-trapping to percolation in the high temperature conductivity of Mn-doped $LaCrO_3$" *Physical Review Letter,* **65** 1383 (1990).

INTERFACE REACTIVITY BETWEEN YTTRIA STABILIZED ZICONIA AND STRONTIUM-LANTHANUM MANGANITES

Monika Backhaus-Ricoult, Michael Badding, Jacqueline Brown, Mike Carson, Earl Sanford, Yves Thibault
Crystalline Materials Research or Characterization Sciences, Science & Technology
Corning Incorporated
Corning NY 14831
USA

ABSTRACT

Formation of pyrochlore at LSM-YSZ interfaces is studied in the temperature range from 700 to 1250°C for various LSM and YSZ compositions. At 900°C, small isolated islands are observed to form in the interfacial plane; continuous scales develop only at much higher temperatures. The very irregular distribution in space and size of the pyrochlore islands formed in the temperature range from 900 - 1250°C reflects high nucleation and growth barriers. Pyrochlore forms more easily for Mn-deficient LSM. For the reaction of LSM with tetragonal 3YSZ, formation of Y,Mn,La-stabilized cubic phase precedes the pyrochlore formation. Single and polycrystalline cubic YSZ exhibit a larger interface coverage than tetragonal YSZ. Intersections of LSM/3YSZ interfaces with 3YSZ grain boundaries are not preferentially decorated with pyrochlore. As a consequence, less pyrochlore is formed for small grain size 3YSZ than for larger size cubic 8YSZ at equivalent reaction conditions.

INTRODUCTION

Yttria stabilized zirconia and $(Sr,La)MO_3$ perovskites with M = Mn, Fe, Co, Ni are commonly used as electrolyte and cathode materials in solid oxide fuel cells (SOFC). For cathodes based on LSM, it is well established that the electrode overpotential is strongly linked to the catalytic efficiency of triple phase boundaries between cathode, electrolyte and gas, where oxygen from the gas phase is incorporated into the electrolyte. The overall oxygen exchange reaction is composed of several parallel and sequential reaction steps, including adsorption of oxygen, its dissociation, electron charge transfer, vacancy ejection from the electrolyte etc. For an efficient cathode operation, the overall reaction rate has to be high. Local chemical composition at surfaces, interfaces and triple phase boundaries play a crucial role for the kinetics of many individual reaction steps. Formation of interfacial reaction products, interdiffusion, segregation and local modifications in the electronic structure at the interface induce modifications in SOFC performance during high temperature processing and operation. A fundamental understanding of the chemical reactivity and associated microstructural changes at cathode-electrolyte interfaces is expected to provide guidelines for the development of more efficient cathodes in high performance SOFC.

Despite its lower electrical conductivity and oxygen exchange rate compared to lanthanum ferrites and cobaltites, substituted lanthanum manganite is the conventional cathode catalyst material of choice in most technical prototypes due to its material stability during high temperature processing and its thermal and chemical compatibility with YSZ. Formation of pyrochlore as reaction product between YSZ and LSM has been widely studied[1-5]. It is commonly agreed that pyrochlore forms only when the manganite reaches a critical depletion in manganese. Therefore, usually an induction period is observed, in which evaporation or out-diffusion of manganese into YSZ occur. Increase in Y- or initial Mn-content in YSZ, Mn-excess in LSM, grain size and decreasing reaction temperature were observed to retard pyrochlore formation.

Since the oxygen ion conductivity of $La_2Zr_2O_7$ is by a factor 1000 smaller than that of YSZ[6], it is evident that any extensive pyrochlore formation limits the SOFC performance. Only at temperatures above 1300°C, a dense scale of pyrochlore forms at the YSZ-LSM interfaces; at lower temperature, isolated islands develop.

Several TEM studies on the nucleation and initial growth mechanisms of pyrochlore and the crystallographic constraints were conducted in simplified model systems: pyrochlore islands were grown on YSZ single crystal surfaces by reactive vapor deposition of La_2O_3 at elevated temperature[7], interfaces in LSM-YSZ composites were investigated after exposure to different thermal treatments and fuel cell operation conditions[8,9] and YSZ substrates with vapor deposited LSM films were characterized[4]. Most literature results cannot be compared because LSM composition, reaction temperature, geometry and material porosity (triple phase boundary density) differ. In addition, most studies are limited to cubic YSZ; only a few data are available for tetragonal partially stabilized zirconia[10,11]. The 3YSZ electrolyte is of interest to the authors due to its superior mechanical properties compared with cubic YSZ, which enable fabrication of a planar bipolar-plate free SOFC design based on thin (20 μm) mechanically flexible 3YSZ electrolyte[12]. Therefore, we conducted a comparative study on the reactivity of single crystalline cubic, polycrystalline cubic and polycrystalline tetragonal YSZ with porous LSM layers of different composition in air in a temperature range of 700 - 1250°C.

EXPERIMENTAL

Reaction couples between polycrystalline 3YSZ, polycrystalline 8YSZ or single crystals of 9.5YSZ and $(Sr_xLa_{1-x})_{1-\delta}MnO_3$ perovskites with $x = 0.1 - 0.2$ and δ ranging from -0.1 to +0.1 were annealed at temperatures between 700 and 1250°C. The polycrystalline zirconia ceramics were obtained by sintering slip-casted Tosoh 3YSZ and 8YSZ powders at 1450°C to full density. 3YSZ ceramics had an average grain size of 300 nm with a narrow grain size distribution. According to X-ray diffraction, they were composed of 97% or more tetragonal phase and only negligible amount of monoclinic phase. Upon long time annealing in air at temperatures between 700°C and 1300°C, no significant phase transformation and increase in grain size were observed. The 8YSZ ceramics sintered to larger grain size, ranging from 2 to 5 μm. Impurity levels in the ceramics were low (chemical spectrographic analysis revealed less than 0.03 wt% SiO_2 and 0.02 wt% Al_2O_3). Single crystals of cubic zirconia were purchased from Coating & Crystal Technology, PA. All ceramics were cut to 10 mm x 10 mm slices. A thick perovskite layer was printed on the zirconia substrates using an ink based on LSM powder with particle size 200 nm from Praxair and organic vehicle. The reaction couples were annealed for 1 to 100 h in air. During firing, the perovskite layer sintered to a porous scale. Thus cathode triple phase boundaries in SOFC were much better simulated than by MBE- or vapor-deposited dense LSM layers. For some of the TEM investigations, higher interface densities were produced in mixed YSZ-LSM composite powder pellets.

Reaction product quality, quantity and distribution were evaluated by X-ray diffraction, scanning electron microscopy (SEM) in a LEO 722 and analytical transmission electron microscopy (TEM) in a JEOL 2000FX equipped with an EDX detector, a LEO EFTEM or the modified VG HB501 STEM of the Cornell CCMR facility.

To evaluate the formation of pyrochlore as reaction product, the reacted samples were etched in concentrated HCl. Thus perovskite was dissolved and interfacial pyrochlore became easily visible on the zirconia substrate. The quantity and distribution of the pyrochlore reaction product was studied by high resolution SEM of the etched surfaces.

Interdiffusion was evaluated on polished cross-sections of the reaction couples by electron microprobe (JEOL JXA-8200) and analytical TEM.

For TEM investigation, cross-sections were wedge-polished to electron transparency and then ion-milled for some minutes. The evolution of the chemical composition across the interfaces was studied by energy dispersive X-ray analysis (TEM/EDX). Chemistry and electronic structure of the interfaces were studied in the Cornell VG HB501 STEM[13] with a cold field emission gun operated at 100 kV. Probe size of 0.22 nm, collection semi-angle of the spectrometer of 15mrad, energy dispersion of 0.7 eV/ch and emission current of 2.3 pA were typical operation parameters. EELS data were acquired by a CCD camera from a Parallel Electron Energy Loss Spectrometer (Gatan PEELS). Spectra were taken for the low loss area and the Sr M, Y M, Zr M, Mn L, O K and La L characteristic absorption edges. Series of line spectra were acquired for the different edges, with the probe being discontinuously stepped across the interface. Spectra were explored manually. The characteristic elemental signals were extracted from the acquired spectra by subtracting the background under the characteristic edges under use of a power law fit. Direct comparison of interfacial spectra with those of the bulk phases was used to identify major changes in the electronic structure at the interfaces. In a "finger print"-type analysis, near edge fine structural details of the interfaces were compared to features of different bulk phase references.

RESULTS

Reaction Products

All studied reaction couples between yttria stabilized zirconia and $(Sr_{0.15}La_{0.85})MnO_3$ exhibited some interfacial reactivity. Type and amount of reaction products differed. Different types of reaction products were observed: pyrochlore $La_2Zr_2O_7$ and solid solutions of cubic zirconia. No significant formation of reaction products was observed for temperatures below 1000°C, and, even at higher temperature, for most LSM compositions long induction times were necessary before pyrochlore covered a substantial part of the interface area.

Distribution Of Pyrochlore

Figure 1 illustrates that, in the temperature range from 900-1250°C, isolated pyrochlore interface precipitates formed in the interface and not a continuous product scale. For reaction couples with single and polycrystalline YSZ, the basic precipitate shape was a platelet of 50 nm in diameter and about 5 nm in thickness. The platelets exhibited rough faceting with very rounded edges. For most particles, the main facets did not lie in the interfacial plane, but were inclined (often perpendicular) to it. At 1000°C, the average pyrochlore platelet size did not increase with time; only very few particles grew within 100 h to a much larger, irregular plate of 1-2 micrometers in size. For perovkites with initial Mn-excess, the density of such large plates remained negligible; for manganese-deficient perovskites, the amount of formed pyrochlore increased proportional to the manganese deficiency. At higher temperatures, in a first reaction stage, the same type of small platelet formed. With increasing reaction time, however, the growth

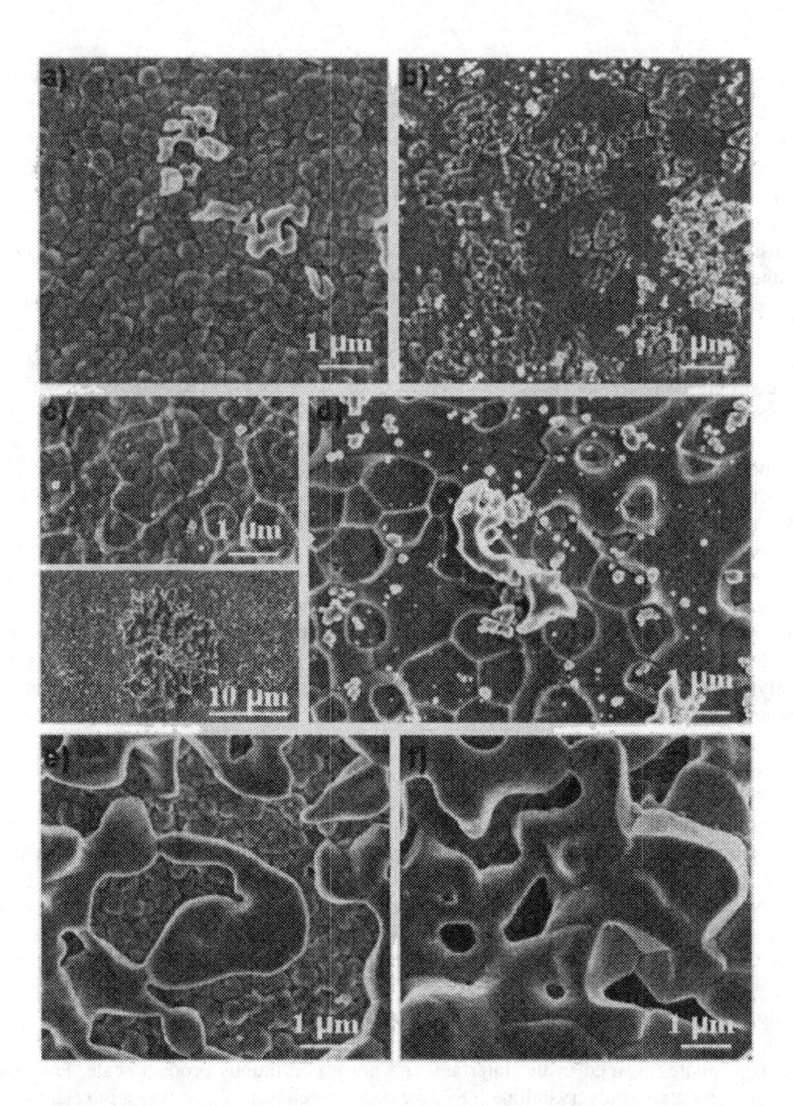

Figure 1: View on the interfacial plane of YSZ//LSM reaction couples after annealing and etching of the LSM phase. Pyrochlore is visible as bright contrast phase. a) 3YSZ//$LSM_{\delta=+0.02}$ after annealing at 1000°C for 100h; b) 8YSZ//$LSM_{\delta=+0.02}$ after annealing at 1000°C for 100h; c) 3YSZ//$LSM_{\delta=+0.02}$ after annealing at 1250°C for 100h; d) 8YSZ//$LSM_{\delta=+0.02}$ after annealing at 1250°C for 100h; e) 3YSZ//$LSM_{\delta=0}$ after annealing at 1250°C for 100h; f) 3YSZ//$LSM_{\delta=-0.05}$ after annealing at 1250°C for 100h.

to large, extended plates became more and more dominant.

Reaction couples with 3YSZ did not show any preferential decoration of interface intersections with YSZ grain boundaries; the occasional precipitates at such locations had the same shape and size as those formed on top of 3YSZ grains. 8YSZ ceramics behaved differently in the reaction couples. While, for short reaction times, intersections of the interface with 8YSZ grain boundaries were not preferentially decorated, for longer reaction times more and more irregular, intergrown agglomerated platelets were found in such grain boundary locations. For the same reaction conditions, we found always more pyrochlore particles in reaction couples with 8YSZ ceramics than for 3YSZ ceramics. In addition, those particles were larger in size. The reaction couples with single crystalline 9.5YSZ confirmed this evolution by showing slightly larger precipitates than reaction couples with 8YSZ.

In all reaction couples, a zirconia rim formed around completely or partially isolated LSM particles on the flat YSZ surface, thus producing a very high curvature in the interfacial plane. This rim was hardly visible at low reaction temperature and for short time, but achieved considerable height for higher temperatures and longer times, Figure 2. The rim height was found to be always larger in reaction couples with 8YSZ ceramics. For all types of ceramics, the rim was composed of Mn-,La-substituted zirconia.

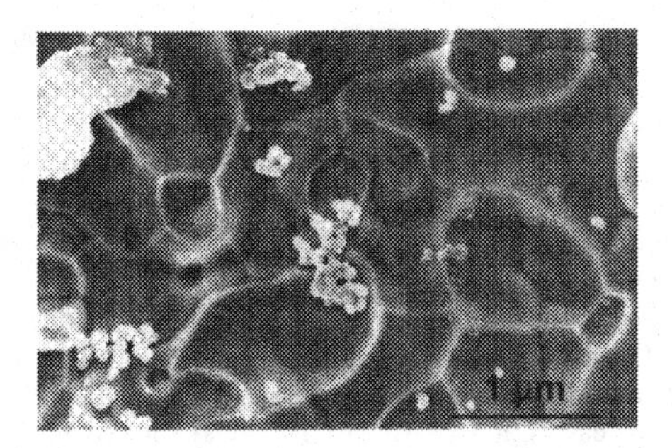

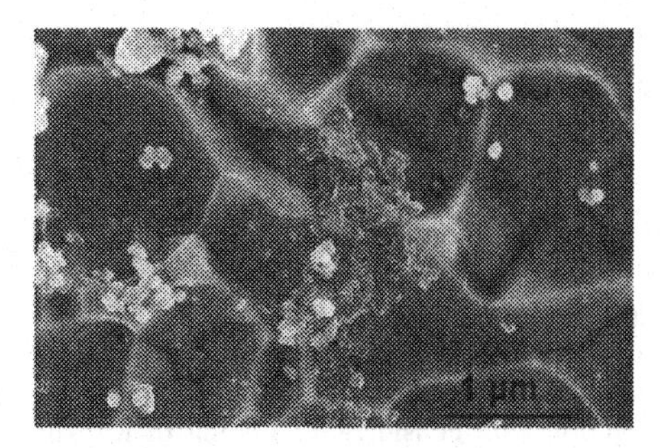

Figure 2: View on the interfacial plane of a) 3YSZ//$LSM_{\delta=+0.02}$ and b) 8YSZ//$LSM_{\delta=+0.02}$ reaction couples after annealing for 20 h at 1250°C and etching of the LSM phase; a large grained YSZ layer has formed with rims that surround the LSM grains (not visible because etched). Bright contrast particles are $La_2Zr_2O_7$.

Interdiffusion Across The Interface

Besides the formation of pyrochlore in the interfacial plane, interdiffusion of various species occurred across the interface and induced changes in the chemical composition in proximity of the interfaces. Since the diffusion of cations in perovskite is much faster than in stabilized zirconia, the main compositional changes were observed in the YSZ ceramic in a thin layer close to the interface. Manganese and lanthanum were found to dissolve from the perovskite into the YSZ ceramic and diffuse into the ceramic (Figures 3 and 4). Bulk diffusion of the cations was found to be rather slow in both, cubic and tetragonal phase compared to the by a factor of 1000 faster grain boundary diffusion. Typical concentration profiles in the 3YSZ and 8YSZ ceramics of reaction couples are shown in Figure 3, illustrating the faster average penetration into 3YSZ due to fast grain boundary diffusion and small grain size. The interdiffusion profiles also revealed that a layer of highly substituted YSZ formed in proximity of the interface in 3YSZ ceramics. Details on diffusion coefficients as function of chemical composition and temperature will be reported elsewhere[14].

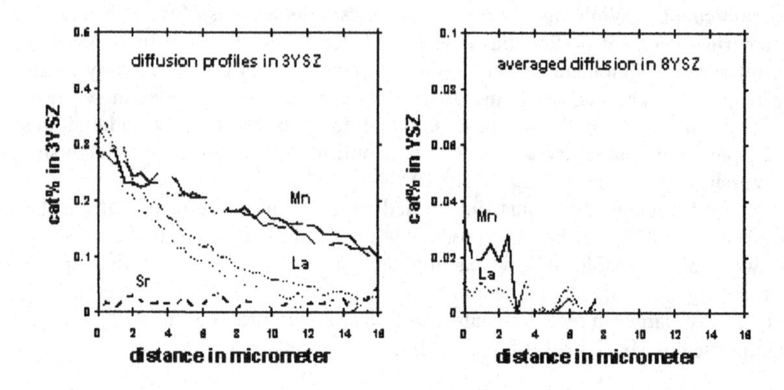

Figure 3: Averaged diffusion profiles of manganese, lanthanum and strontium in the zirconia phase after annealing at 1250°C for 20 h. Reaction couples of LSM with 3YSZ and 8YSZ are compared.

Local Chemistry And Electronic Structure Of YSZ-LSM-Oxygen Triple Phase Boundaries

The local chemistry at pyrochlore-free YSZ-LSM-gas triple phase boundaries was studied by spatially resolved TEM/EDX and TEM/EELS in 3YSZ-LSM composites reacted at 1250°C for 2 h in air. Typical scans across the interface of the O, Mn, La, Y and Zr characteristic absorption edges are presented in Figure 4. The manganese and lanthanum scans show that a zirconia solid solution with considerable amount of Mn (up to 14 cat%) and La (up to 10 cat%) formed in direct contact with the perovskite. Dissolved strontium levels remained under the detection limit by EDX or EELS in the TEM. The LSM grains usually showed a small depletion in manganese compared to the starting perovskite composition.

The near edge fine structures of the characteristic absorption edges of the different elements revealed that the preponderant valence states of dissolved manganese in the zirconia solid solution was Mn^{2+} and for lanthanum La^{3+}. Compared to grains far away from the interfaces, perovskite in contact with YSZ did not show any modification in crystal structure or valence state of manganese and lanthanum.

At the interface, the following characteristic features were repeatedly observed: An enhanced Y-edge intensity reflected a significant segregation of yttrium to the interface. The L_3,L_2 peak energies and L_3:L_2 peak height ratios in the interfacial manganese spectrum reflected a mixture of Mn^{2+}, Mn^{3+} and Mn^{4+}; especially high Mn^{2+}concentrations were observed in the interfacial plane, Figure 4. A shoulder in the interfacial O-K edge on its high energy side indicated an oxygen excess at the interface, with excess oxygen being bonded to Mn^{2+}.

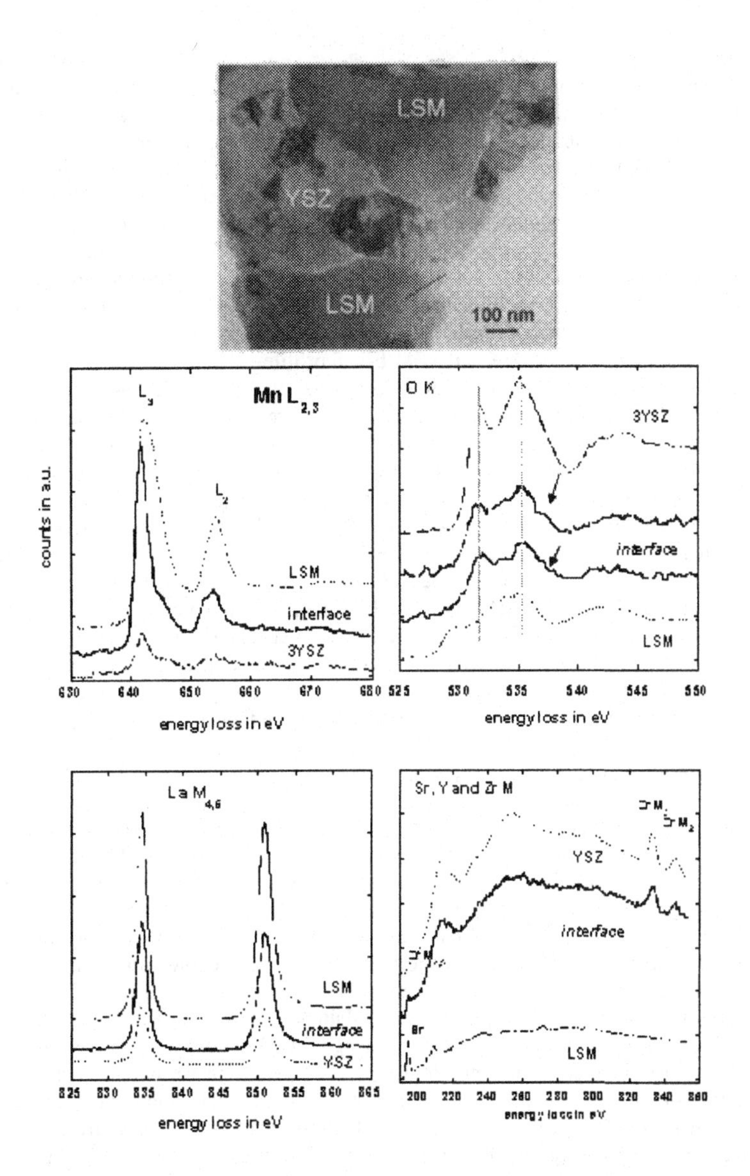

Figure 4: TEM micrograph of a LSM-3YSZ reaction mixture after annealing in air at 1250°C for 2 h together with a series of ELNES spectra of the Mn-, O-, La-, Sr-, Y- and Zr- absorption edges that were acquired in the YSZ-phase, in the LSM-phase and at the interface location, while stepping the electron beam across the LSM-3YSZ interface.

DISCUSSION

Thermodynamic models predict that pyrochlore forms as reaction product between yttria stabilized zirconia and strontium lanthanum manganite for manganese activities below a critical threshold value, whereas, above that threshold, Mn-saturated cubic zirconia and LSM can coexist[1]. This suggests that interdiffusion across the interface always takes place in LSM-YSZ reaction couples. Since diffusion of manganese is much faster and its solubility higher than that of lanthanum or strontium, zirconia enriches mainly in manganese. We noticed considerable interdiffusion in reaction couples with single and polycrystalline YSZ. While interdiffusion for cubic 8YSZ and 9.5YSZ lead only to a partial substitution of zirconium and yttrium by manganese and also at a lower level by lanthanum, for 3YSZ, it lead in addition to the formation of an intermediate layer of cubic zirconia. We identified the cubic phase through its high cation substitution in microprobe, TEM/EDX and TEM/EELS profiles.

For reaction temperatures under 1300°C, formation of pyrochlore and even the transformation of tetragonal zirconia into the cubic phase require high activation energies and therefore occur only at preferential nucleation sites. As a consequence, continuous product layers have difficulty to develop. We observed irregularly distributed isolated pyrochlore islands in the interface. Not only their formation seemed to be associated to a high energy barrier, but also their growth: Only few pyrochlore plates were observed to grow to larger size. Several reasons can be found for such behavior: the diffusional transport alimenting the reaction can be slow, the crystallographic phase transformation can be inhibited due to large associated reaction stresses and slow stress relaxation or defect relaxation can be sluggish. In the present case, the pyrochlore platelet morphology suggests that the pyrochlore interface energy is anisotropic. From a simple energy balance, spreading of the reaction product in the interfacial plane is expected to minimize the interface energy and the total energy of the system. Such energy minimization, however, is not met for the growth of the small pyrochlore platelets, since their large facets do not lie in the interfacial plane.

Our observations of annealed reaction couples revealed that a small rim formed at the LSM-YSZ interfaces. Such rim formation is usually observed when, upon annealing, a spherical particle (that can be considered as a solid "drop") on a flat surface tends to reach through diffusional transport the equilibrium morphology for coexistent gas, solid and liquid phases. The equilibrium "wetting" angle of the YSZ substrate by an LSM "drop" in air is given by the force equilibrium at the triple phase boundary between the two solids, YSZ and gas and LSM and gas. The initial contact angle between YSZ substrate and LSM grains does not correspond to the equilibrium state, therefore, diffusional transport (plastic deformation including cation and anion transport) occurs and helps to o establish the triple phase boundary equilibrium geometry. While for liquid drops the entire drop can easily adopt a flattened shape and wet the flat surface, for solid "drops" or particles, such macroscopic change in morphology of an entire grain is extremely slow. Instead, local diffusional transport in the triple phase boundary area leads to restricted changes in the local morphology, such as rim or groove formation. In the present case, the formation of a rim of doped YSZ suggests that transport of yttrium and zirconium cations occurs by bulk diffusion through LSM, while oxygen is transported through the gas phase or on the surface.

It is possible that the catalytic activity for the oxygen reduction at triple phase boundaries is improved in porous composite cathodes by such a growing rim, because continuously new, clean triple phase boundary is created while the rim advances. The growth of the YSZ rims may also explain why most small pyrochlore platelets do not have their large facets in the macroscopic planar interface plane. The interfaces between LSM grain and YSZ in the thin rim constitute preferential sites for pyrochlore nucleation because of easier stress relaxation in the thin zirconia layer of the rim.

TEM studies on the formation of pyrochlore islands on (001) YSZ single crystal surfaces from vapor deposited La_2O_3 at elevated temperature[7] demonstrated that pyrochlore islands grow on free YSZ surfaces in a cube on cube orientation as pyramids to approximately 500 nm diameter and 50 nm in height. Those pyramids were usually composed of several sectors with slightly rotated variants. In case of perfectly flat zirconia surfaces, the island growth seemed to be controlled by the formation of dislocations at the island corners, where growth stresses were highest, followed by rapid glide of the dislocations. For terraced surfaces, it is reported that surface steps were preferentially decorated because stresses of the growing islands easier relaxed by deformation of the pit rim. The latter mechanism was expected by Hesse et al. to be in certain cases slower, suggesting then that rough surfaces were less reactive than smooth YSZ surfaces.

That thesis was made for pyrochlore islands growing on free surfaces; pyrochlore formation at interfaces between two solid phases has to meet more restrictions. As a result, some of the experimental observations differ. For the solid state reaction, YSZ grain boundary grooves in the LSM/YSZ interface were not preferentially decorated and small grain size YSZ ceramics (larger surface curvature and higher step density) formed less pyrochlore islands. These observations indicate that for the solid state reaction accommodation of crystallographic structure and growth stresses are not only coupled to the YSZ phase, but also to the perovskite phase.

The finding that YSZ grain boundary intersections with the interface are not preferentially decorated together with the fact that an additional energy barrier has to be overcome to transform tetragonal doped zirconia into the cubic phase provide a lower chemical reactivity of 3YSZ ceramics with LSM. This finding is new and in contrast to observations made at higher temperature (1400 and 1500°C)[12]. It shows that 3YSZ ceramics compared to 8YSZ ceramics have not only the better mechanical properties, but also a lower reactivity with LSM, thus suggesting them more suited as electrolyte material in YSZ-based SOFC.

CONCLUSIONS

The experimental study of YSZ-LSM interface reactivity comparing single crystalline 10YSZ, polycrystalline cubic 8YSZ and nanocrystalline tetragonal 3YSZ ceramics showed that for all investigated temperatures and LSM chemical compositions, quantity and size of pyrochlore product islands remained slightly smaller for the tetragonal 3YSZ ceramic. Further on, TEM/EELS studies have shown that a manganese-rich thin bulk diffusion layer always forms on the zirconia side at the 3YSZ-LSM interface. While bulk diffusion of manganese, lanthanum and strontium into YSZ is found to be very slow, grain boundary diffusion is by a factor of 1000 faster. Manganese enrichment in the grain boundary core is found to reach up to 14 cat%, lanthanum enrichment up to 12 cat%.

ACKNOWLEDGEMENTS

This work made use of the ultra-high vacuum scanning transmission electron microscopy and microscopy facilities of the Cornell Center for Materials Research (CCMR) with support from the National Science Foundation Materials Research and Engineering Centers (MRSEC) program (DMR-0079992). The authors want to especially thank M. Thomas, CCMR, for his help.

REFERENCES

[1] "High temperature solid oxide fuel cells", ed. S.C. Singhal, K. Kendall, Elsevier, Oxford, 2004

[2] A. Mitterdorfer, L. Gauckler, "$La_2Zr_2O_7$ formation and oxygen reduction kinetics of the $La_{0.85}Sr_{0.15}MnO_3$, O_2(gas)/YSZ system", *Solid State Ionics* **111**[3] 185-218 (1998)

[3] L. Kindermann, D. Das, D. Bahadur, R. Weiss, H. Nickel, K. Hilpert, "Chemical interactions between La-Sr-Mn-Fe-O based perovskites and yttria stabilized zirconia", *Journal of the American Ceramic Society* **80**[4] 909-915 (1997)

[4] T. Horita, T. Tsunoda, K. Yamaji, N. Sakai, T. Kato, H. Yokokawa, "Microstructures and oxygen diffusion at the $LaMnO_3$ film/yttria stabilized zirconia interface", *Solid State Ionics* **152/153** 439-446 (2002)

[5] M. Mori, T. Abe, H. Itoh, O. Yamamoto, G.Q. Shen, Y. Takeda, N. Imanishi, "Reaction mechanisms between lanthanum manganite and yttria doped zirconia", *Solid State Ionics* **123**[1-4] 113-119 (1999)

[6] "Materials for fuel cells", *Annual Review of Materials Research* **33** (2003)

[7] C.J. Lu, S. Senz, D. Hesse, "The impact of YSZ surface steps on structure and morphology of $La_2Zr_2O_7$ islands growing on YSZ (100) surfaces by vapour-solid reaction", *Surface Science* **515**[2/3] 507-516 (2002)

[8] D.M. Tricker, W.M. Stobbs, "The microstructure of solid oxide fuel cells and related metal-oxide interfaces", *High temperature electrochemical behavior of fast ion and mixed conductors* **14**, 453-460 (1993), editors F.W. Poulsen et al., Risoe National Laboratory, Roskilde

[9] C. Clausen, C. Bagger, J.B. Bilde-Soerenen, A. Horsewell, "Microstructural and microchemical characterisation of the interface between $La_{0.85}Sr_{0.15}MnO_3$ and Y_2O_3 stabilized zironia", *Solid State Ionics* **70/71**[1] 59-64 (1994)

[10] S.P. Jiang, J.-P. Zhang, K. Foeger, "Chemical interaction between 3mol% yttria zirconia and strontium-doped lanthanum manganite", *Journal of the European Ceramic Society* **23**[4] 1865-1873 (2003)

[11] J.A.M. Van Roosmalen, E.H.P. Cordfunke, "Chemical reactivity and interdiffusion of (La,Sr)MnO_3 and (Zr,Y)O_2 solid oxide fuel cell and electrolyte materials", *Solid State Ionics* **52**[4] 303-312 (1992)

[12] M. Badding, J. Brown, T. Ketcham, D., St.Julien, R. Wusirika, "High performance solid electrolyte fuel cells" US6623881 B2, US-patent (1998)

[13] http://www.CCMR.Cornell.edu/facilities/ high resolution STEM facility

[14] M. Backhaus-Ricoult, Y. Thibault, "Comparison of cation diffusion in doped 8YSZ and 3YSZ ceramics", to be published

ELASTIC PROPERTIES, BIAXIAL STRENGTH AND FRACTURE TOUGHNESS OF NICKEL-BASED ANODE MATERIALS FOR SOLID OXIDE FUEL CELLS

Miladin Radovic, Edgar Lara-Curzio, Beth Armstrong and Claudia Walls
Metals and Ceramics, Oak Ridge National Laboratory
Oak Ridge, TN 37831-6069

ABSTRACT

NiO-YSZ composites are widely used as precursors for Ni-YSZ anode materials for solid oxide fuel cells (SOFCs) with YSZ electrolytes. Prior to operation, the anode material is reduced in hydrogen to obtain a cermet comprised of metallic Ni and YSZ. The reduction process is accompanied by changes in the microstructure and mechanical properties of the Ni-based anodes, which needs to be understood and optimized to maximize the reliability of SOFCs. In this study the elastic properties, biaxial strength and fracture toughness of Ni-based anode materials were determined at room temperature as a function of initial porosity before and after full reduction in hydrogen. The elastic properties were determined by impulse excitation, while biaxial strength and fracture toughness were determined using the ring-on-ring and double torsion test methods, respectively. It was found that the magnitude of Young's and shear moduli decreases significantly with porosity and that the magnitude of the elastic moduli decreases almost by 50% due to H_2-induced reduction. The characteristic strength of the distributions of biaxial strengths of Ni-based anode materials decreased with increasing porosity, and it was always found to be lower (by as much as 50%) than that of the initial NiO-YSZ material. The decrease in the magnitude of the biaxial strength after reduction was attributed mainly to the increase in porosity of the material. Conversely, the fracture toughness of fully reduced Ni-based anodes was found to be significantly higher than that of unreduced anodes, which results from the ductile behavior of nickel.

INTRODUCTION

Solid Oxide Fuel Cells (SOFCs) have been intensively studied during the past decade as highly efficient and pollution-free energy sources. Research related to SOFCs have been generally focused on electrochemical, thermal and microstructural properties of the different materials in SOFCs, as well as on electrochemical performance and processing of different SOFC geometries [1,2]. However, reliability of SOFC does not depend only on the chemical and electrochemical stability of its components but also on the capability of the SOFC components to withstand mechanical stresses that arise during processing and service.

The room temperature elastic properties of NiO-YSZ containing 75mol% of NiO have been reported by Selcuk and Atkinson[3,4] and Radovic et al.[5,6]. These studies showed that room temperature Young's and shear moduli decrease with increase in porosity. Young's modulus of NiO-YSZ decreases by ≈50% from ≈200 GPa to 110 GPa with increase in porosity form zero to 23 vol% after hydrogen reduction. More recently, Radovic and Lara-Curzio[6] reported on the changes of elastic properties of Ni-based anode as a function of extent of reduction process. They showed that Young's and shear moduli of Ni-based anode with initial concentration of 75mol% NiO and initial porosity of 23vol% decreases approximately 45% after hydrogen reduction mainly due to increase in porosity. The Young's modulus of ≈60 GPa determined for fully reduced Ni-based anode is considerably lower then that of YSZ electrolyte (≈200 GPa)[1]. Such a discrepancy in elastic properties between adjacent layers has a significant effect on the stress state in the multilayered SOFC. Not much has been reported on the strength and fracture toughens of Ni-based anodes. Atkinson and Selcuk[3] reported on ambient temperature biaxial strength of the NiO-YSZ anode with 75mol% of NiO. They determined the caharacteristic strength of 187 MPa and Weibull modulus of 11.8 for the samples of unknown porosity. Radovic et al.[5] reported a decrease in room temperature biaxial strength of NiO-YSZ precursor for anode with porosity. Besides these studies, there are very few reports on the strength and fracture properties of the Ni-based anodes. However, the effect of porosity, chemical composition, environment and temperature on mechanical properties of Ni-based anode is still not understood.

In this paper we report the room temperature elastic properties, biaxial strength and fracture toughness of Ni-based SOFC anode with 75 mol% of NiO as a function of porosity before and after hydrogen reduction. Elastic moduli of the unreduced and reduced Ni-based anode material were determined by Impulse Excitation technique. Biaxial strength was determined using the ring-on-ring configuration while fracture toughness of the anode samples was obtained by the double torsion method.

EXPERIMENTAL PROCEDURE

The Ni-based anode materials used in this study were prepared from a powder mixture of NiO (J.T.Baker*, Phillipsburg, NJ) and ZrO_2 stabilized with 8 mol% Y_2O_3 (TOSOH Corp.*, Grove City, OH). The powder mixture contained 75mol% of NiO and 25mol% of Y_2O_3 stabilized ZrO_2 (YSZ for shorter). Different amounts of organic pore former (rice starch, ICN Biomedicals*, Inc Irvine, CA) were added to the powder mixture to obtain samples with different levels of porosity. Green samples were prepared by tape casting ≈250 μm thick single layers. Four green tapes were subsequently laminated to make samples of different thickness. Discs for determination of elastic properties and biaxial strength with nominal diameter of 25.4 mm were hot-knifed from the assembled green tapes and sintered at 1400 °C in air for 2 hours. Rectangular samples for double torsion testing were cut from tapes previously sintered also at 1400 °C in air for 2 hours. Approximately one half of the samples were reduced in 4% H_2 and 96% Ar gas mixture at 1000 °C for 30 minutes. The relative porosity of each sample before and after reduction was determined using the standard alcohol immersion method[7].

The elastic properties of un-reduced and fully reduced anode samples, namely Young's and shear moduli were determined by impulse excitation (IE) technique[8] using the commercially available Buz-o-sonic* software program (BuzMac Software, Glendale, WI). Disc-shaped anode specimens supported by a foam material on its nodal lines were excited by a light mechanical impulse. A microphone, located in the vicinity of the sample is used to transmit sound vibrations to the signal-processing unit. The fundamental resonance frequencies, in both torsional and flexural mode were identified, which in turn can be used to calculate values of Young's and Shear moduli using standard equations for disc-shaped samples[8].

It is usual practice to determine the strength of the SOFC materials using concentric ring-on-ring flexural-loading configuration[9,10,11] because components of planar SOFC stacks are mainly exposed to biaxial states of stress that results from the thermal expansion mismatch between electrolyte, anode and cathode materials and temperature gradients during operation. In the ring-on-ring configuration the disk-shaped test specimens with diameter ≈25 mm were spaced concentrically between a loading ring with diameter of 5.5 mm and a supporting ring with diameter of 20 mm. Load was applied to the samples trough the loading ring at a constant cross head displacement rate of 1 mm/min until failure. In all characterized samples failure originated

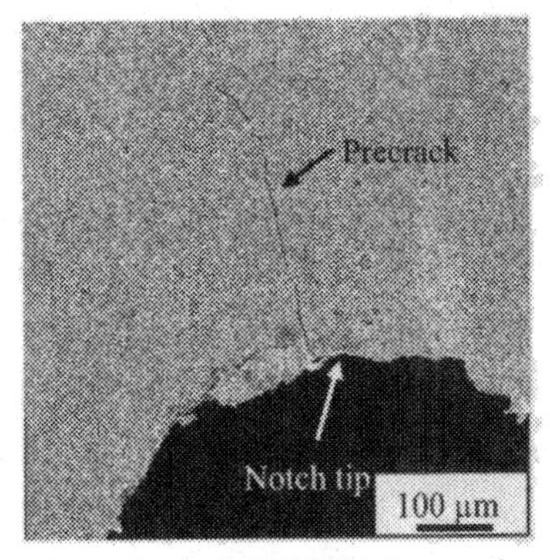

Fig. 1. Typical FESEM micrograph of the the precrack at the notch tip of double torsion test specimen. Reduced Ni-based anode with 31.8 vol% porosity.

within perimeter of the loading ring and was in agreement with guidance provided by the ASTM C1499 standard test method[9].

The fracture toughness, K_{IC}, of Ni-based anode material was determined using the double torsion method[12,13,14]. The tests were conducted using rectangular double torsion samples (20 mm wide, 40 mm long and ≈1 mm thick). Initial notches 1 mm wide and 12.5 mm long were introduced into one side of the sample using a circular diamond blade. The notch tip was machined such that the thickness of the specimen at the notch tip tapered from very thin to the full thickness of the sample. This facilitates the formation of a sharp precrack at relatively low loads. The notched test specimen was precracked by loading the sample at rate of 0.02 mm/min. Typical sharp precracks initiated at the notch tip on the lower side of the sample is shown on Fig. 1. Precracked samples were loaded at a high loading rate of 2 mm/min to cause fast fracture. The fracture toughness of Ni-based anode samples was calculated using the equation for the plain stress conditions[12,13].

The microstructure and fracture surfaces of Ni-based anode samples were analyzed using a Field Emission Scanning Electron Microscope (FESEM) Hitachi S4700* before and after hydrogen reduction.

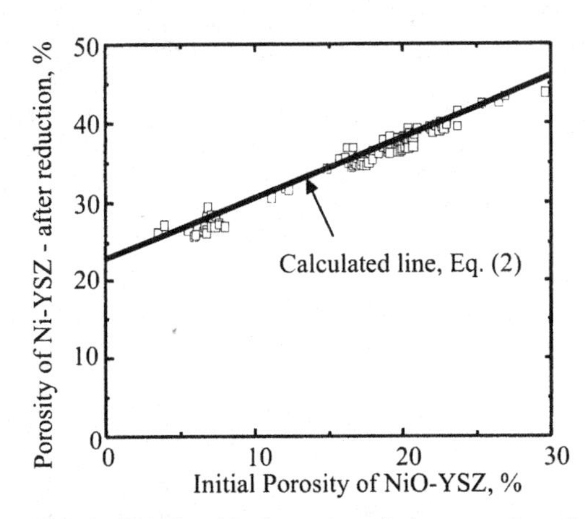

Fig. 2. Relationship between relative porosity of the reduced anode samples and initial porosity of the samples before reduction. Open square symbols represent experimental results determined by the alcohol immersion method, while the solid line is representation of Eq.(2).

EXPERIMENTAL RESULTS AND DISCUSSION

Changes in Porosity

The porosity of the anode samples change as a result of hydrogen reduction as shown in Fig. 2 where relative porosity of the reduced anode is plotted versus initial porosity of the samples. The open square symbols in Fig. 2 represent results obtained experimentally by alcohol immersion method. The increase in porosity is expected since the specific volume of metallic Ni is significantly smaller than that of NiO. If we assume that the decrease of the overall volume of anode samples after reduction is negligible, then the following expression can be derived for the porosity of the anode sample after reduction, *p*, as the function of initial porosity before reduction, p_0:

$$p = p_0 + (1 - p_0) \cdot \overline{m}_{NiO}^0 \cdot \left[\frac{1}{\rho_{NiO}} - \frac{1}{\rho_{Ni}} + \frac{m_O}{m_{NiO}} \cdot \frac{1}{\rho_{Ni}} \right] \quad (1)$$

where ρ_{Ni} and ρ_{NiO} are the average density of Ni and density of NiO, respectively. $\overline{m}_{NiO}^0$ is the initial weight fraction of NiO in the NiO-YSZ composite. For the examined NiO-YSZ composite the initial weight fraction of NiO was $\overline{m}_{NiO}^0 = 0.587$. For ρ_{Ni}=8.88 g/cm^3 and ρ_{NiO}=6.67 g/cm^3 Eq.(1) yields:

$$p = 0.228 + 0.772 p_0 \quad (2)$$

Equation (2) has also been plotted in Figure 2 as which is in good agreement with the experimental results.

Table 1. The best fit values of zero-porosity moduli, (E_0 and G_0), the porosity dependence constants (b_E and b_G) and coreleation cofficient (R^2) for NiO-YSZ and corresponding NI-YSZ determined using exponential and CSM models.

		Material	
		NiO-YSZ	Ni-YSZ
Exponential Model	E_0, GPa	210.8±1.4	212.1±1.31
	b_E	2.7±0.04	3.16±0.03
	R^2	0.985	0.995
	G_0, GPa	80.4±0.5	79.5±0.6
	b_G	2.6±0.04	2.95±0.03
	R^2	0.984	0.992
CSM	E_0, GPa	209.5±1.4	211.4±0.9
	b_E	0.46±0.05	0.79±0.02
	R^2	0.984	0.998
	G_0, GPa	79.4±0.6	78.2±0.4
	b_G	0.33±0.05	0.53±0.02
	R^2	0.982	0.997

Elastic properties

Figure 3 shows Young's (E) and shear (G) moduli determined by IE as a function of porosity for unreduced (NiO-YSZ) and reduced (Ni-YSZ) anode materials at ambient temperature. Among the large number of empirical and theoretical models that have been proposed to express the dependence of the magnitude of Young's and Shear moduli on the volume fraction of porosity, the following two expressions[3] were used to fit the experimental modulus-porosity data in the present work:

Exponential

$$M = M_{fd} \exp(-b_M p') \quad (3)$$

Composite Spheres Model:

$$M = M_{fd} \frac{(1-p')^2}{1 + b_M p'} \quad (4)$$

where M_{fd} is are elastic moduli (E or G) of the fully dense material, b_M is a porosity dependence constant and p' is fractional porosity. The modulus-porosity relationships obtained using exponential and CSM models are also shown in Fig. 2., while the zero porosity moduli values (E_0 and G_0) and porosity dependent constants (b_E and b_G) are listed in Table 1. The high values of R^2 and small standard errors (Table 1) for both, exponential and CSM model indicate that both models provides equally good fits. The zero-porosity elastic moduli values and porosity dependence coefficients for NiO-YSZ are in good agreement with previously published values by Selcuk and Atkinson[3] for 75mol%NiO-YSZ. On the other hand, the similarity of the zero-

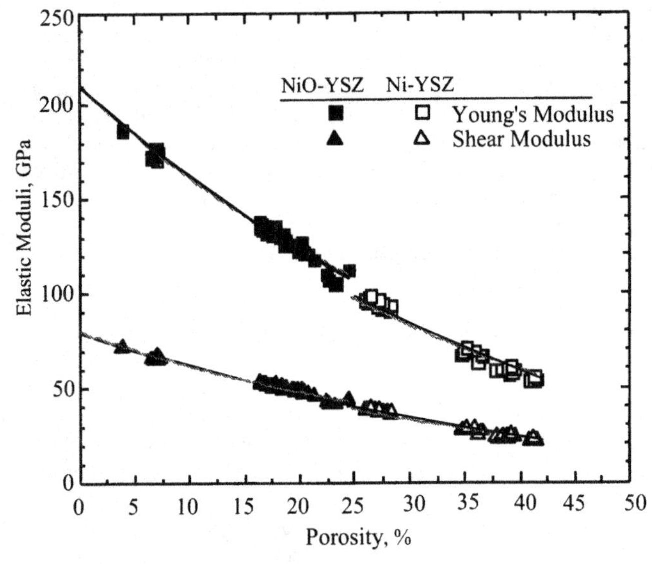

Fig. 2. Young's (E) and shear (G) moduli of unreduced and reduced Ni-based anode as a function of porosity. The results of fitting using exponential and CSM model are represented as gray and black solid lines, respectively.

moduli and porosity dependence coefficients determined for unreduced and reduced anode indicates that the decrease in the elastic moduli during the reduction of NiO-YSZ anode is almost exclusively due to increase in porosity, as previously reported by Radovic and Lara-Curzio[6].

Biaxial strength

The ambient temperature biaxial strength results for Ni-based anodes were analyzed using a two-parameter Weibull distribution.:

$$f(\sigma) = 1 - exp\left[\left(-\sigma/\sigma_0\right)^m\right] \tag{5}$$

Table 2. Results of biaxial testing at room temperature for 22vol% porous NiO-YSZ and 40vol% porous Ni-YSZ samples with different number of laminated layers.

	Porosity*, vol%	Biaxial Strength*, MPa	Characteristic biaxial strength**, MPa	Weibull modulus**	Sample size
NiO-YSZ	6.6±0.8	127.4±17.1	134.6 (125.6,143.6)	8.6 (5.6,12.1)	15
	17.8±0.8	88.9±23.7	97.6 (86.4,110.2)	4.3 (2.9,6.4)	15
	19.8±1.1	86.0±23.7	92.1 (84.5,100.0)	6.8 (4.3,9.8)	15
	21.9±1.8	86.0±21.5	95.4 (83.3,109.1)	4.0 (2.6,6.0)	15
Ni-YSZ	27.4±0.5	107.1±19.7	115.2 (104.2,126.2)	6.1 (3.9,8.67)	15
	36.1±0.4	74.3±10.9	79.1 (73.2,85.4)	7.0 (4.8,10.1)	15
	36.9±1.0	67.9±13.9	73.5 (66.1,81.8)	5.0 (3.5,7.2)	15
	39.8±1.6	50.7±12.0	55.42 (49.2,62.5)	4.5 (3.1,6.5)	15

* (Average value) ± (standard deviation) ** Inside brackets is the 95% confidence interval.

where $f(\sigma)$ is the probability density function of biaxial strength, σ is the failure stress of a specimen, σ_0 is the scale parameter or Weibull characteristic strength and m is known as shape parameter or Weibull modulus†. Weibull distribution parameters were estimated using maximum likelihood technique and the Kaolan-Meier ranking method, while 95% two-sided confidence bounds were calculated using Fisher's matrix method. The parameters of the Weibull distribution and average strengths for Ni-based anodes with different porosity are listed in Table 2 and plotted as a function of porosity in Fig. 3 for Ni-based anode before and after hydrogen reduction. Fig. 3 indicates that the characteristic strength decreases with porosity for both, NiO-YSZ and corresponding Ni-YSZ specimens.

The dependence of strength on porosity has been described by:

$$\sigma = \sigma_0 \exp(-b_\sigma p') \quad (6)$$

where σ_0 is strength of nonporous structure. Rice[15,16] showed that the Eq. (6) can be successfully applied for describing strength-porosity relationship for a wide range of different materials and that the decrease of strength with porosity is proportional to the decrease of elastic moduli with porosity according to the minimum solid area model. The fit of characteristic strength-porosity data using Eq. (6) results in σ_0=158.7±8.7 MPa and b_σ=2.58±0.34 with a correlation factor R^2=0.962 for unreduced anode material, and σ_0=473.4 ±104.3 MPa and b_σ=5.12±0.67 with a correlation factor R^2=0.965 for reduced anode material. The high R^2 value suggests that strength-porosity behavior can be well described using Eq. (6). Results shown in Fig. 3 indicate that the characteristic biaxial strength of anode material decreases due to hydrogen reduction. On the other hand, for the same porosity value the characteristic strength of reduced Ni-YSZ anode material is higher that that of unreduced NiO-YSZ. Thus, it can be concluded that the observed reduction in biaxial strength as a result of reduction is for the most part caused by the increase in porosity in the sample.

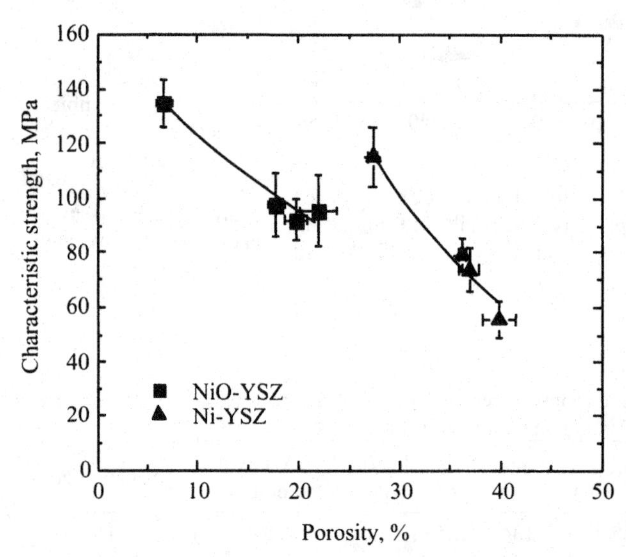

Fig. 3. Characteristic biaxial strength determined form Weibull distribution for NiO-YSZ and Ni-YSZ samples with different porosity. Solid lines represent results of fitting strength-porosity data using Eq. 6.

† The Weibull analysis was carried out using commercially available software Weibull++[1] (ReliaSoft, Tucson, AZ)

Fig. 4 Typical failure origins: (a) cluster of NiO particles close to tensile surfaces, NiO-YSZ with ≈6.6 vol% porousity; (b) large cavity close to tensile surface, NiO-YSZ with ≈6.6 vol% porousity; (c) cluster of YSZ grains near the tensile surface, NiO-YSZ with ≈6.6 vol% porousity; (d) surface dimple, NiO-YSZ with 21,9 vol% porousity; (e) cluster of YSZ grains on the tensile surface, NiO-YSZ with 21,9 vol% porousity; (f) a cavity close to tenisle surface, Ni-YSZ with ≈27.3 vol% porousity; (g) cluster of Ni grains on tensile surface, Ni-YSZ with ≈27.3 vol% porousity.

Fracture surfaces after biaxial testing were analyzed using a Field Emission Scanning Electron Microscope. The high level of porosity of the samples usually makes it difficult to identify fracture origins because most of the characteristic fractographic features such as mirror, mist and hackle regions are not apparent on the fracture surface. However, some typical fracture origins in NiO-YSZ and Ni-YSZ samples were identified as illustrated in Fig. 4. These include clusters of NiO particles (Fig. 4a), large cavities (Fig. 4b), YSZ agglomerates (Fig. 4c and 4e) and dimples (Fig. 4d) close that were identified as fracture origins in NiO-YSZ specimens. In the case of Ni-YSZ samples, large cavities (Fig. 4f), clusters of Ni grains (Fig. 4g) and YSZ agglomerates (not shown here) were usually identified as origins of the fracture.

Fracture toughness

Average values of fracture toughness, K_{IC}, determined using the double torsion method at ambient temperature are listed in Table 3 and plotted in Fig. 4 for NiO-YSZ and Ni-YSZ specimens with different porosity values. It was found that fracture toughness of both, NiO-YSZ and Ni-YSZ decrease with increasing porosity. The results in Fig. 5 also indicate that the reduction process is accompanied by an increase in fracture toughness due to the conversion of NiO into Ni. The results in Fig. 5 can be expressed the using following relationship:

$$K_{IC} = K_{IC0} \exp(-b_K p') \quad (7)$$

where K_{IC0} is the fracture toughness of the nonporous structure. The fitting of the model to the results yields values of K_{IC0}=2.54±0.38 MPam$^{1/2}$ and b_K=2.51±0.83 for unreduced anode material and K_{IC0}=7.52±0.93 and b_K=3.03±0.36 for reduced material. Field Emission Scanning Electron Microscopy analysis of precracks and fractures surfaces (not shown here) indicated that transgranular fracture is the predominant mechanism for crack propagation in NiO-YSZ. However, Fractographic analysis revealed plastic deformation of the Ni-metallic phase in Ni-YSZ. Thus, plastic deformation of Ni grains ahead of the crack tip must contribute significantly to the higher fracture toughness of Ni-YSZ material comparing to that of NiO-YSZ.

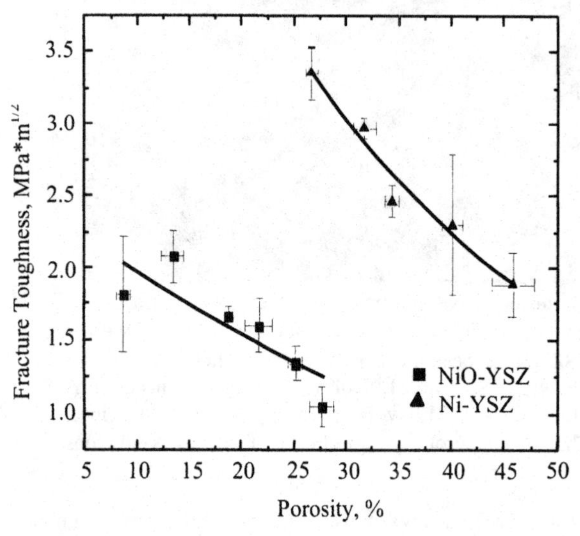

Fig. 4. Fracture toughness of NiO-YSZ and Ni-YSZ as a function of porosity.

CONCLUSIONS

Porosity and hydrogen reduction were found to have a significant effect on Young's and shear moduli, biaxial strength and fracture toughness of Ni-based anodes. Hydrogen reduction of NiO-YSZ to Ni-YSZ leads not only to changes in the chemical composition, but also to increase in porosity of the anode material. Young's and shear moduli of both, NiO-YSZ and Ni-YSZ material were found to decrease with porosity. The decrease in elastic moduli after reduction can be mostly related to the increase in porosity of the samples. Biaxial strength and fracture toughness of both, unreduced and reduced Ni-based anodes were also found to decrease with increasing porosity. Hydrogen reduction of the NiO-YSZ precursor

Table 3. Fracture toughness of NiO-YSZ and corresponding Ni-YSZ after reduction in hydrogen. Fracture toughness was determined from double torsion tests.

NiO-YSZ (before reduction)			Ni-YSZ (after reduction)		
Porosity*, vol%	Fracture toughness K_{IC}*, MPa	Sample size	Porosity*, vol%	Fracture toughness K_{IC}*, MPa	Sample size
8.8±0.7	1.8±0.4	4	26.7±0.5	3.4±0.2	5
13.5±1.1	2.1±0.2	10	31.8±1.1	3.0±0.1	4
18.8±0.2	1.7±0.1	2	34.4±0.6	2.5±0.1	5
21.8±1.3	1.6±0.2	5	40.1±1.0	2.3±0.5	5
25.3±0.7	1.3±0.1	2	-	-	-
27.6±1.1	1.1±0.1	5	45.9±2.0	1.9±0.2	5

* (Average value) ± (standard deviation)

causes a decrease in equibiaxial strength. However, fracture toughness of the anode samples increases after reduction in hydrogen, despite the significant increase in porosity during reduction. The considerable increase of fracture toughness can be explained by formation of the ductile Ni-metal phase during the reduction. The property-porosity trends are in good agreement with the minimum solid area model.

ACKNOWLEDGMENTS
This research work was sponsored by the US Department of Energy, Office of Fossil Energy, SECA Core Technology Program at ORNL under Contract DE-AC05-00OR22725 with UT-Battelle, LLC. The authors are grateful for the support of NETL program managers Wayne Surdoval, Travis Shultz and Donald Collins. The authors are indebted to Shawn Reeves of ORNL for help with thermogravimetric measurements.

REFERENCES

[1]Ming N. Q. and Takahashi T., "Science and Technology of Ceramic Fuel Cells", Elsevier, Amsterdam (1995

[2]Zhu W.Z. and Deevi S.C., "A review on the Status of Anode Materials for Solid Oxide Fuel Cells", Mater. Sci. Eng. **A326** (2003) 228

[3]Selcuk A. and Atkinson A., "Elastic Properties of Ceramic Oxides Used inSolid Oxide Fuel Cells (SOFC)", *J. Euro. Ceram. Soc.* **17** (1997) 1523

[4]Atkinson A. and Selcuk A., "Mechanical Behavior of Ceramic Oxygen Ion-Conducting Membranes", Solid State Ionics 134 (2000) 59

[5]Radovic M., Lara-Curzio E., Armstrong B. and Walls C., "Effect of Thickness and Porosity on the Mechanical Properties of Planar Components for Solid Oxide Fuel Cells at Ambient and Elevated Temperatures", in "27th Cocoa Beach Conference on Advanced Ceramics and Composites Ceramics" edited by Kriven W.M. and Lin H.T., Engineering and Science Proceedings, Vol. 24, The American Ceramic Society (2003) 329

[6]Radovic M. and Lara-Curzio E., "Change of Elastic Properties of Nickel-based Anodes for Solid Oxide Fuel Cells as a Function of the Fraction of Reduced NiO", J. Am. Ceram. Soc., in print.

[7]ASTM standard C20

[8]ASTM C1259

[9]ASTM standard C1499

[10]Morrell R. "Biaxial Flexural Strength Testing of Ceramics Materials", Measurement Good Practice Guide No. 12, National Physical Laboratory, Teddington, United Kingdom (1998)

[11]Salem J. and Powers L., "Guidelines for the Testing of the Plates", in "27th Cocoa Beach Conference on Advanced Ceramics and Composites Ceramics" edited by Kriven W. M. and Lin H. T., Engineering and Science Proceedings, Vol. 24, The American Ceramic Society (2003).

[12]Fuller E. R. Jr, "An Evaluation of Double-Torsion Testing – Analysis" in *Fracture Mechanics Applied to Brittle Materials*, ASTM Special Technical Publication No 678 Edited by Freiman S. W., ASTM, Philadelphia (1997)

[13]Pletka B.J, Fuller E. R. Jr and Koepke B. G. "An Evaluation of Double-Torsion Testing – Analysis", ibid 16

[14]Tait R. B., Fry P. R. and Garrett G. G., "Review and Evaluation of the Double –Torsion Techniques for Fracture Toughness and Fatigue Testing of Brittle Materials", Exp. Mech. 14-22 (1987) pp. 14

[15]R. W. Rice, "Evaluation and Extension of Physical Property-Porosity Models Based on Minimum Solid Area", J. Mater. Sci. **31** (1996) 102

[16]R. W. Rice, "Comparison of Physical Property-Porosity Behavior with Minimum Solid Area Models", J. Mater. Sci. **31** (1996) 1509

[•] Certain commercial equipment, instruments, or materials are identified in this paper in order to specify the experimental procedure adequately. Such identification is not intended to imply recommendation or endorsement by Oak Ridge National Laboratory, nor is it intended to imply that the materials or equipment identified are necessarily the best available for the purpose.

EFFECT OF HYDROGEN REDUCTION ON THE MICROSTRUCTURE AND ELASTIC PROPERTIES OF Ni-BASED ANODES FOR SOFCs

Miladin Radovic, Edgar Lara-Curzio, Beth Armstrong, Peter Tortorelli and Larry Walker
Metals and Ceramics, Oak Ridge National Laboratory
Oak Ridge, TN 37831-6069

ABSTRACT

One route for the synthesis of solid-oxide fuel cells incorporating Ni-based anodes involves co-sintering in air of the electrolyte with a NiO-YSZ precursor to a Ni-YSZ anode. Prior to the operation of the cell it is necessary to convert NiO into metallic Ni by hydrogen-reduction. The reduction of NiO into Ni is characterized by significant volumetric changes, changes in porosity and hence, elastic properties and strength. These changes in turn will modify the state of residual stresses in the cell and impact its reliability. In this study, the kinetics of hydrogen reduction (using a gas mixture of 4%H_2-96%Ar) of 23vol% porous 75mol%NiO/YSZ anode materials was investigated by thermogravimetry between 600°C and 800°C. In addition, samples were reduced at 800°C for different periods of time to monitor the evolution of structural changes as a function of fraction of reduced NiO. The kinetics of reduction were found to exhibit two stages: At all temperatures the fraction of reduced NiO was found to increase linearly with time until nearly 70-80% of NiO was reduced, and the rate of reduction was found to increase with temperature according to an Arrhenius law with an activation energy of 25.2 kJ/mol. Optical and scanning electron microscopy, and electron microprobe chemical analysis indicate that the first stage of the reduction process is associated with the displacement of the reduction front across the thickness of the sample, whereas the second stage, which occurs at a much slower rate involves further reduction of NiO behind the reduction front. Young's and shear moduli of Ni-based anodes were determined by Resonant Ultrasound Spectroscopy and Impulse Excitation as a function of fraction of reduced NiO. It was found that elastic moduli decrease with extent of the reduction reaction predominately due to increase in porosity.

INTRODUCTION

Solid Oxide Fuel Cells (SOFCs) are electrochemical devices that convert the chemical energy of a fuel (e.g. – hydrogen, natural gas, reformed gasoline or diesel) directly into electrical energy. The building blocks of SOFCs include a fully-dense electrolyte layer capable of conducting oxygen ions, which is embedded between electrically conducting anode and cathode layers. Currently a cermet consisting of interconnecting and interpenetrating Ni and Y_2O_3 stabilized ZrO_2 (YSZ) is widely used as the anode material for many SOFCs. The porous Ni-YSZ anodes have good electronic conductivity, stability, catalytic properties and compatibility with other constituents of SOFC[1,2]. Also, thermal expansion behavior of Ni-YSZ cermets can be tailored to match that of other components of the SOFC[3,4]. High porosity (up to 50 vol%) of Ni-based anodes is required to obtain high external current densities and to enable hydrogen to infiltrate the cermet and H_2O to escape[5]. While Ni metal has a vital role as a catalyst for electrochemical reaction on the anode side of the SOFC and as electrical current conductor[5], YSZ have mainly mechanical function as support for Ni particles and prevent them from sintering during the service.

A common fabrication procedure of SOFC includes co-sintering of electrolyte, porous cathode and porous NiO-YSZ precursor for Ni-YSZ anode. Prior to the operation of the cell it is necessary to reduce NiO into metallic Ni by hydrogen-reduction. During the reduction process the chemistry, microstructure and properties of the anode change. The reduction of NiO into Ni is characterized by significant volumetric changes, changes in porosity and hence, elastic properties, thermal expansion coefficients and strength.

Thermal expansion studies[7-9] have shown that both Ni-YSZ and NiO-YSZ have a higher coefficient of thermal expansion than YSZ electrolytes and that the coefficient of thermal expansion increases with the amount of NiO/Ni content in the anode. It has also been reported[8] that the coefficient of thermal expansion of porous anodes does not change significantly after reduction of NiO into Ni, at least for anodes with NiO content up to 85 mol%. For example, coefficient of thermal expansion for NiO content of 75mol% was determined to be $12.7 \times 10^{-6}\,K^{-1}$ before reduction and $11.7 \times 10^{-6}\,K^{-1}$ after reduction in hydrogen. Itoh[9] and co-workers reported the shrinkage of the Ni-based anode form 3 to 17% after reducing and holding the sample in hydrogen for 300 h. They also showed that the microstructure of the Ni-based anode has a strong effect on overall shrinkage during reduction. The changes of elastic moduli, coefficients of thermal expansion and volumetric changes of Ni-based anode during initial hydrogen reduction modify the state of residual stresses in the cell and impact its reliability[6]. It is therefore expected that volume changes and changes in elastic properties should have a major effect on stress redistribution in multilayered cells during reduction. Since stresses introduced during the reduction step would have a significant effect on the reliability and durability of SOFC, it is necessary to understand the kinetics and structural changes associated with this process, as well as a changes in elastic properties.

EXPERIMENTAL PROCEDURE

A mixture of 75mol% of NiO (J.T.Baker*, Phillipsburg, NJ) and YSZ (8mol%YSZ, TOSOH Corp.*, Grove City, OH) powders with addition of 30vol% of organic pore former (rice starch, ICN Biomedicals*, Inc Irvine, CA) was used to tape cast green tapes of porous NiO-YSZ precursor for anode. Green samples were laminated into ≈1 mm thick tapes and disc-shaped specimens (nominal diameter 25.4 mm) were hot-knifed from the assembled green tapes. Specimens were sintered at 1450 °C in air for 2 hours. The relative porosity of the samples after sintering was found to be 23.5±0.6 vol% as determined by alcohol immersion[11] (Archimedes method).

The individual samples were reduced in a thermo-gravimetric unit (Cahn 1000*) with a detection limit of 0.1 mg. The sample mass was monitored continuously during reduction using an automatic data acquisition system. Samples were placed on a quartz stand that was hung by a Pt suspension wire in the thermobalance and introduced into a quartz reaction tube. The sample was heated under a constant flow of argon (99.999% purity) to the reaction temperature. No mass changes were observed during the heating period. After reaching thermal equilibrium, the gas was changed to a mixture of 4%H_2+96%Ar (min 99.95% pure) which was flowed at a constant rate of 100 cc/min for the duration of the reduction process. Reduction runs were carried out for different times at 800 °C to monitor microstructural changes and determine elastic properties as a function of reduction time. Also, some samples were fully reduced at 600, 650, 700, 750 and 800°C to investigate the effect of temperature on reduction kinetics. The relative porosity of partially and fully reduced samples was also determined by alcohol immersion[11].

Elastic moduli, namely Young's modulus, E, and shear modulus, G of the examined samples were determined by Impulse Excitation, IE and Resonant Ultrasound Spectroscopy, RUS at room temperature. IE allows for the determination of E and G from the resonant frequencies of a mechanically excited sample[12]. According to this test procedure a disc-shaped specimen is supported by a foam material on its nodal lines and then excited by a light mechanical impulse. A microphone, located in the vicinity of the sample is used to transmit sound vibrations to the signal processing unit. The fundamental resonance frequencies, in both flexural and torsional mode, are identified, which in turn can be used to calculate values of E and G for the sample of known dimensions and mass. In this study the commercially-available Buzz-o-sonic* (BuzMac Software, Glendale, WI) program was used to determine the fundamental frequencies and calculate the elastic moduli of the samples. On the other hand, RUS is a technique that can be used for determining elastic constants of solids from the resonant spectrum of mechanical resonance for a sample of known geometry, dimensions and mass[13]. The disc-shaped specimens were supported by three piezoelectric transducers. One transducer (transmitting transducer) is used to generate an elastic wave of constant amplitude and varying frequency, whereas the other two transducers are used to detect resonance. An approximate spectrum is calculated from the

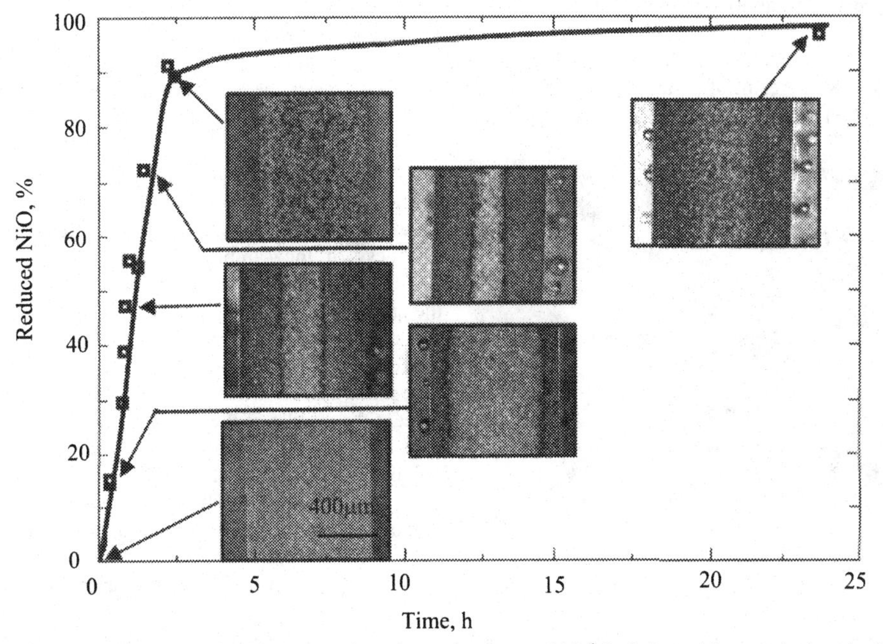

Fig. 1. Fraction of reduced NiO vs. time for reduction at 800 °C. Selected but typical optical micrographs show the microstructure of samples with different fractions of reduced NiO. Dark gray layers on the both sides of the sample correspond to the reduced part of the samples.

known sample dimensions, density and a set of "guessed" elastic constants. A multidimensional algorithm (Quasar International*, Albuquerque, NM) to minimize root-mean-square (RMS) error between the measured and calculated resonant peaks enables the estimation of the elastic constants of the solid from the single frequency scan.

RESULTS AND DISCUSSION

Kinetics of Hydrogen Reduction

The reduction of the NiO by hydrogen can be represented by following reaction:

$$NiO(s) + H_2(g) \Rightarrow Ni(s) + H_2O(g) \quad (1)$$

The mass change recorded by the thermo-gravimetric unit during the reduction process was converted to the fraction of reduced NiO, r, namely the ratio of the instant mass change ΔW_t, to the theoretical final mass change, ΔW_{th}. The later was calculated for every sample using the following equation and assuming that no other reaction except reaction (1) occurred during the reduction process:

$$\Delta W_{th} = (m_O / m_{NiO}) \cdot W_{NiO} \quad (2)$$

where W_{NiO} is the initial mass of NiO in the sample. m_O and m_{NiO} are the molecular mass of oxygen (≈16 amu) and of NiO (≈74.7 amu), respectively.

Figure 1 shows the fraction of reduced NiO, r as a function of reduction time. Open squares in Fig. 1 represent results from interrupted reduction tests at 800°C are, while the solid line in Fig.1 is the result of an uninterrupted test. These results suggests the existence of two stages. In the first stage, the reduction is nearly linear with time until ≈85% of NiO has been reduced. Beyond this transition point the amount of mass loss with time decreases gradually. Optical micrographs shown in Fig. 1 illustrate the development of the reduced layer across the thickness of the samples during reduction at 800°C, where dark gray layers on both sides of the samples are reduced parts, while light gray areas represent unreduced parts. The micrographs in Fig. 1. suggest that the transition occurs when the "reaction fronts" meet at the middle of the sample. However, this does not mean that all NiO is completely reduced within the gray area.

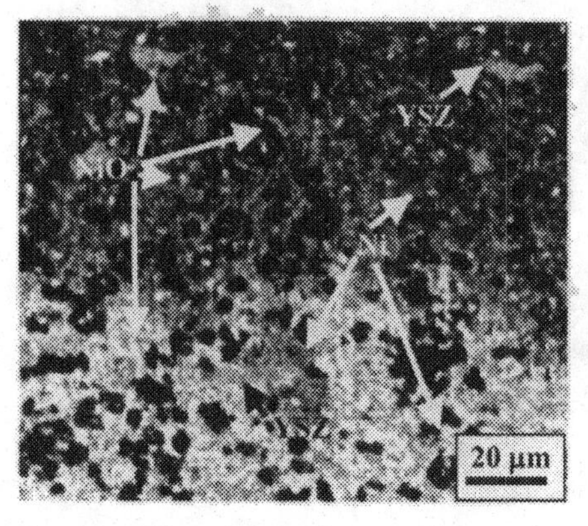

Fig. 2. Elemental surface analysis of a reaction front in partially reduced anode sample obtained using JEOL 8200* Electron Superprobe. Gray is YSZ, white is NiO and dark gray is Ni-metal. Black areas are pores.

Typical results of surface elemental analysis obtained using scanning electron microscopy with wavelength dispersive spectrometry (WDS) are shown in Fig. 2. It was found that the boundary between reacted and unreacted zone is rather diffusive. NiO grains (white in Fig. 2) were found in predominately reduced areas, while Ni-metal (dark gray) was also found in the predominately unreduced zone. The shape of reduction curve (Fig.1) and existence of diffusive reaction zone (Fig. 2) suggest that reduction kinetics can be explained by grain model developed by Szekely, Evans and co-workers[14-17] to describe hydrogen reduction of pure NiO. According to that model, the gaseous reactant (i.e. hydrogen) diffuses from the surface of

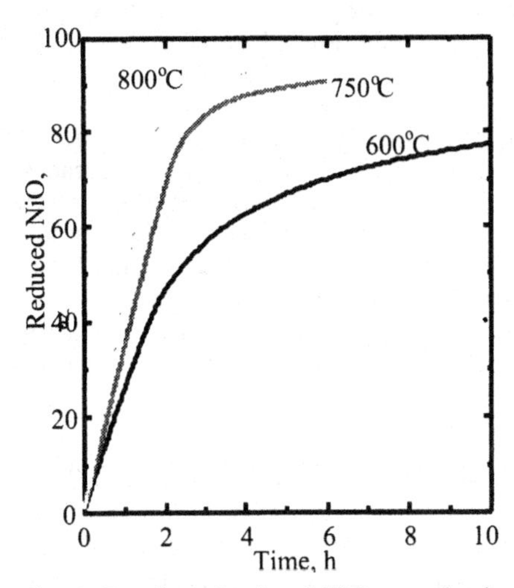

Fig. 3. Fraction of reduced NiO vs. reduction time at different temperatures.

the sample through interstices and reacts with the spherical particles. Then, the reaction between the gas and the solid phase proceeds in accordance with the "shrinking core" model. Gas diffuses through the product layer (Ni) in each grain and reacts at the spherical reaction interface, while the generated product gases (H_2O) diffuse back through the solid product and between the grains into the bulk gas steam. This results in a diffusive reaction zone, i.e. progressively lesser amount of reaction with the depth of penetration of the gas stream. According to the grain model the reaction rate is controlled by the rate of chemical reduction and/or diffusion of the gaseous phase through the pores. The present results suggest that this is the case during the extended first stage, when reaction kinetics are linear, but that diffusion through a tortuous gas path or in the solid state may be controlling thereafter as the amount of further reduction per unit time continually decreases.

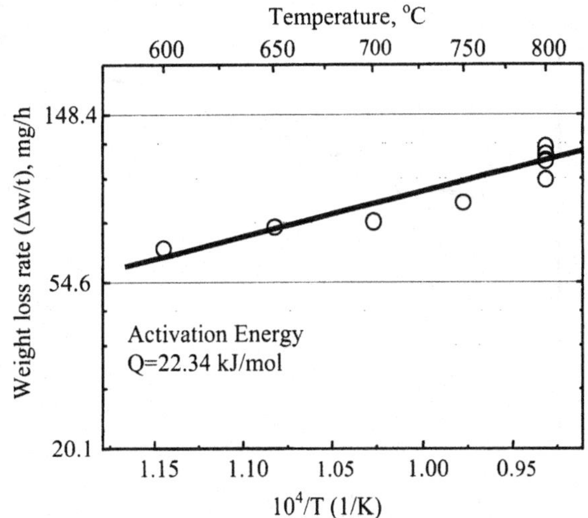

Fig. 4. Arrhenius plot for the reduction of Ni-based anode in 4% and 96% argon gas mixture.

Figure 3 shows selected but typical curves showing the fraction of reduced NiO as a function of reduction time at different temperatures. The shape of those curves indicate that the transition point in reduction kinetics depends on temperature suggesting that reduction is probably controlled by both, the rate of the chemical reaction and the diffusion of the gases through the pores. The initial weight loss rates obtained from the reduction of NiO-YSZ samples in hydrogen are plotted vs. the reciprocial of the reduction temperatures in Fig. 4. The activation energy of 22.3 kJ/mol calculated from the slope of the straight line on Fig. 4 is close to the activation energies of 21.8 and 18.0 kJ/mol that have been reported for the reduction of pure NiO in the same temperature range[18]. This suggests that the presence of YSZ phase in the examined samples does not have a significant effect on the mechanism of NiO reduction.

Porosity Changes during Hydrogen Reduction

The relative porosity of test specimens, p, is plotted in Fig. 5 as a function of the fraction of reduced NiO, r. This trend is expected since the specific volume of metallic Ni is significantly smaller than that of NiO. Scanning - electron micrographs of specimens (not shown here) also confirmed the significant increase in porosity with fraction of reduced NiO. In this case pores that were formed due to shrinkage of NiO particles during their reduction into Ni are much smaller (≈ 1-2 μm) than those initially present in NiO-YSZ samples (10-15 μm).

If it is assumed that the decrease in the overall volume of the samples after reduction is negligible then the following expression can be derived for the changes in porosity as a function of fraction of reduced NiO:

$$p = p_0 + \rho_0 \quad \overline{m}^0_{NiO} \left(\frac{1}{\rho_{NiO}} \quad \frac{1}{\rho_{Ni}} + \frac{m_O}{m_{NiO}} \quad \frac{1}{\rho_{Ni}} \right) r \qquad (3)$$

where p_0, ρ_0, ρ_{Ni} and ρ_{NiO} are the average initial porosity, initial density of the samples, density of Ni and density of NiO, respectively, and $\overline{m}^0_{NiO}$ is the initial weight fraction of NiO in the NiO-YSZ. For the examined NiO-YSZ composite the initial weight fraction of NiO was $\overline{m}^0_{NiO}$ = 0.587. For p_0=0.23, ρ_0=4.83 g/cm^3, ρ_{Ni}=8.88 g/cm^3 and ρ_{NiO}=6.67 g/cm^3 Eq.(3) yields:

$$p \quad 0.23 \quad 0.174 \; r \qquad (4)$$

Equation 4, which has also been plotted in Figure 6, confirms the good agreement between experimentally obtained changes of relative porosity after reduction of NiO.

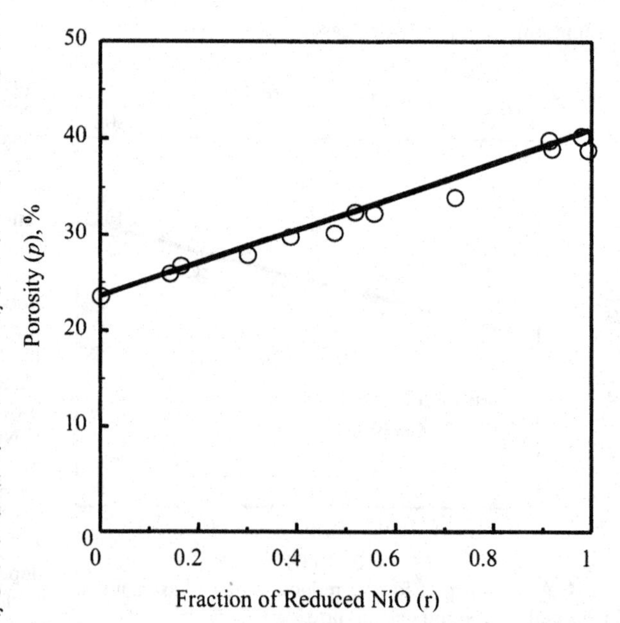

Fig. 5. Porosity of partially reduced samples as a function of fraction of reduced NiO. Solid line corresponds to Eq. (4). Each datapoint represents one measurement per sample.

Changes of Elastic Moduli with Reduction

Young's (E) and shear (G) moduli of samples that were determined at room temperature by RUS (open circles in Fig. 6) and IE (closed triangles in Fig. 6) plotted as a function of fraction of reduced NiO indicate a good agreement between values determined from RUS and from IE. The values of Young's and shear moduli of unreduced NiO-YSZ samples were determined to be in the range of 93.6-103.3 GPa and 36.2-40.3 GPa, respectively. According to the results reported by Selcuk and Atkinson[19] the magnitudes of Young's and shear moduli of NiO-YSZ anode samples with 23.5 vol% porosity should be in the range of 105.0-116.2 GPa and 40.2-44.7 GPa respectively, depending on the model employed to fit elastic moduli vs. porosity data. These results can be considered to be in good agreement with the results presented here, especially if we take into account that Selcuk and Atkinson[19] investigated anode materials with a maximum porosity that is much lower than the porosity of the samples examined in this study.

The decrease in elastic moduli with the fraction of reduced NiO (Fig. 6) can result form: (1) changes in the composition of the anode and/or (2) increase in porosity. To quantify the effects of composition changes on the magnitude of the elastic moduli, the elastic modulus of the anode samples can be expressed as[20]:

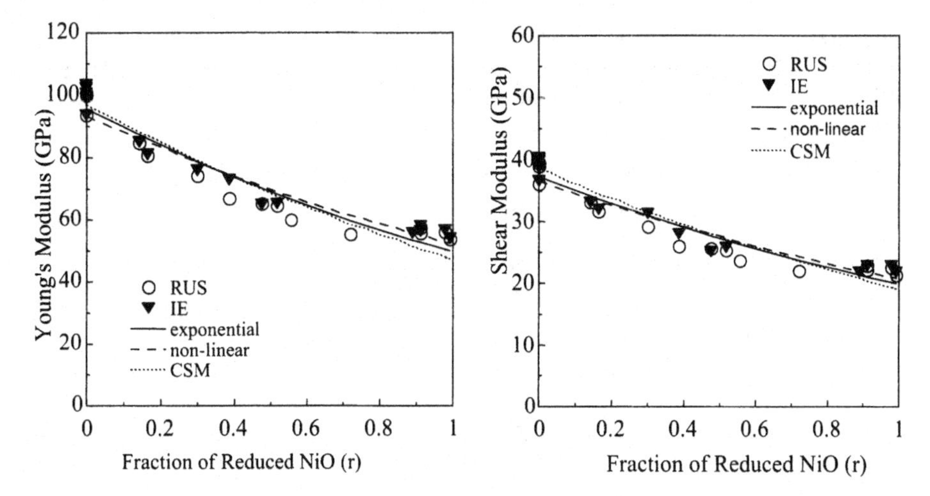

Fig. 6. Young's (left) and shear (right) moduli of anode material as a function of fraction of reduced NiO determined using RUS (open circles) and IE (closed triangles) at room temperature. Also included curves associated with exponential, non-linear and composite sphere models. Each datapoint represents one measurement per sample.

$$M(r) = M_{fd}(r) \cdot M'(p) = M_{fd}(r) \cdot M'(p(r)) \tag{5}$$

where $M(r)$ is the effective modulus (E or G) of the porous NiO-Ni-YSZ composite that is a function of the fraction of reduced NiO. M_{fd} is the effective elastic modulus of fully dense NiO-Ni-YSZ, which is also a function of reduced NiO. $M'(p)$ provides a functional dependence of the elastic properties on porosity. Since the last term in Equation (5) can be expressed as a function of r for the known relationship between the porosity and fraction of reduced NiO (Equation (4)), the changes in the magnitude of the elastic moduli of anodes after reduction can be expressed as a function of only the fraction of reduced NiO.

The effective moduli of the fully dense NiO-Ni-YSZ composite, i.e. the first term in Eq. (5), can be calculated using the rule of mixtures as:

$$M_{fd} = M_{YSZ} \cdot \overline{V}_{YSZ} + M_{NiO} \cdot \overline{V}_{NiO} + M_{Ni} \cdot \overline{V}_{Ni} \tag{6}$$

where M_{YZS}, M_{NiO} and M_{Ni} are the elastic moduli (E or G) of YSZ, NiO and Ni respectively. $\overline{V}_{YSZ}$, $\overline{V}_{NiO}$ and $\overline{V}_{Ni}$ are the volume fractions of YSZ, NiO and Ni respectively. Since , $\overline{V}_{YSZ}$, $\overline{V}_{NiO}$ and $\overline{V}_{Ni}$ are functions of the fraction of reduced NiO, Eq.(6) can be rearranged to obtain:

$$M_{fd}(r) = \frac{M_{YSZ} \cdot \overline{V}^0_{YSZ} + M_{NiO} \cdot (1-r) \cdot \overline{V}^0_{NiO} + M_{NiO} \cdot r \cdot \overline{m}^0_{NiO} \cdot \dfrac{\rho^0_{fd} \cdot m_{Ni}}{\rho_{Ni} \cdot m_{NiO}}}{1 - r \cdot \overline{m}^0_{NiO} \cdot \rho^0_{fd} \cdot \left(\dfrac{1}{\rho_{NiO}} - \dfrac{m_{Ni}}{\rho_{Ni} \cdot m_{NiO}} \right)} \tag{7}$$

where $\overline{V}^0_{YSZ}$ and $\overline{V}^0_{NiO}$ are the initial volume fraction of YSZ and NiO in the NiO-YSZ samples; ρ^0_{fd} is the theoretical density of fully-dense NiO-YSZ; m_{Ni} is the atomic weight of Ni, and the rest of the symbols have previously been described. For moduli of fully dense Ni (E=200 GPa and G=77 GPa) [21], NiO (E=220 GPa and G=84 GPa) [21], 8mol%YSZ (E= 220 GPa and G=83.3 GPa)[18] and for $\overline{V}^0_{YSZ}$=0.443, $\overline{V}^0_{NiO}$=0.557, ρ^0_{fd}= 6.238 g/cm^3 and m_{Ni}=58.7 amu, Equation (7) can be rewritten as:

$$E_{fd} = (220 - 56.82 \cdot r)/(1 - 0.228 \cdot r) \tag{8a}$$

$$G_{fd} = (83.69 - 24.46 \cdot f)/(1 - 0.228 \cdot r) \tag{8b}$$

The following models were used to express the dependence of the magnitude of Young's and shear moduli on the volume fraction of porosity[18]:

$$\text{Exponential: } M = M_{fd} \cdot \exp(-b_M \cdot p) \tag{9}$$

$$\text{Non-linear: } M = M_{fd} \cdot \left[1 - \frac{b_M \cdot p}{1 + (b_M - 1) \cdot p} \right] \tag{10}$$

$$\text{Composite sphere model (CSM): } M = M_{fd} \frac{(1-p)^2}{1 + b_M \cdot p} \tag{11}$$

where M_{fd} is elastic moduli of the fully dense material and b_M is an empirical constant. The iterative least squares fitting of models expressed by Eqs. (8)-(11) to the experimental data result in are b_E and b_G and coefficient of correlation, R^2 that are listed in Table 1. The results of iterative least squares fitting to the experimental data are also shown in Fig. 6. It was found that the three models used to express the dependence of the elastic modulus on porosity (Equations 9-11) yielded very similar results.

Using the composite sphere model, the following relations are obtained to express the dependence of the elastic moduli on the fraction of reduced NiO in the anode materials examined:

$$E = \frac{(220 - 56.82 \cdot r)}{(1 - 0.228 \cdot r)} \cdot \frac{(1 - (0.2355 + 0.174 \cdot r))^2}{1 + b_E \cdot (0.2355 + 0.174 \cdot r)} \tag{12a}$$

$$G = \frac{(83.69 - 24.46 \cdot r)}{(1 - 0.228 \cdot r)} \cdot \frac{(1 - (0.2355 + 0.174 \cdot r))^2}{1 + b_G \cdot (0.2355 + 0.174 \cdot r)} \quad (12b)$$

Table 1. Results of least squares fitting of experimental data

Model	b_E	R^2	b_G	R^2
Exponential Model	3.54±0.04	0.933	3.42±0.04	0.920
Non-linear Model	4.40±0.08	0.923	4.17±0.08	0.923
CSM	1.38±0.06	0.918	1.16±0.06	0.887

The results of this study indicate that the increase in porosity after reduction of NiO into Ni is the major reason for the significant decrease in the magnitude of the elastic moduli of the material. Since the elastic moduli of fully dense Ni, NiO and YSZ are comparable to each other, the changes in the chemical composition of the anode material after reduction only have a slight effect on the magnitude of the effective elastic moduli of the anode material. Equations (7) and (8) show that the decrease in the elastic moduli of a hypothetical fully dense material that is fully reduced and remains fully dense is only ≈4%. Thus, the observed decrease in the elastic properties as a result of reduction of the anode material of ≈45 % can be attributed to the significant increase in porosity with reduction.

SUMMARY

The reduction of 75mol%NiO/YSZ anode material for SOFC with initial porosity of 23vol% in 4%H_2+96%Ar gas mixture and 600-800°C temperature range was studied, as well as corresponding changes in elastic moduli. It was found that reduction kinetics exhibit two stages: the first stage, which takes place through most of the reduction process, occurs almost at a constant rate; during the second stage the rate of reduction decrease significantly with time. The results presented here suggest that the grain model can be used to describe the reduction that is controlled by the rates of chemical reaction and diffusion of the gaseous phases through pores over most of the time while transport through the reduced phase and or tortuous gas paths plays a role in the latter stages. The value of the activation energy, ≈22 kJ/mol was obtained for the first stage of the reduction that is in good agreement with published values for the reduction of pure NiO. The porosity of the samples was found to increase, while elastic moduli were found to decrease with the fraction of reduced NiO. An expression was derived to predict the volume fraction of porosity as a function of the fraction of reduced NiO and it was found that predictions were in good agreement with experimental results. The significant decrease in the magnitude of the elastic moduli of 75mol%NiO-YSZ after reduction in hydrogen was found to be mostly related to increases in the porosity of the sample. By combining the derived relationship between porosity and fraction of reduced NiO with models for the dependence of the elastic properties on porosity, it was possible to obtain a direct correlation between the magnitude of the elastic moduli and the fraction of reduced NiO.

ACKNOWLEDGMENTS

This research work was sponsored by the US Department of Energy, Office of Fossil Energy, SECA Core Technology Program at ORNL under Contract DE-AC05-00OR22725 with UT-Battelle, LLC. The authors are grateful for the support of NETL program managers Wayne Surdoval, Travis Shultz and Donald Collins. The authors are indebted to Shawn Reeves of ORNL for help with thermogravimetric measurements.

REFERENCES

[1]Ming N. Q. and Takahashi T., "Science and Technology of Ceramic Fuel Cells", Elsevier, Amsterdam (1995)

[2]Zhu W.Z. and Deevi S.C., "A review on the Status of Anode Materials for Solid Oxide Fuel Cells", *Mater. Sci. Eng.* A326, 228 (2003)

[3] Muller A.C., Pei B., Weber A. and Ivers-Tiffe E., "Properties of Ni/YSZ Cermet Depending on Their Microstructure", *HTMC IUPAC* July (2000)

[4]Mori M., Yamamoto T. and Itoh H., "Thermal Expansion of Nickel-Zirconia Anodes in Solid Oxide Fuel Cells During Fabrication and Operation", *J. Electrochem. Soc.* **145** (1998) 1374

[5]Morgensen M. and Skaarup S., "Kinetic and Geometry Aspects of Solid Oxide Fuel Cell Electrodes", *Solid State Ionics* 86-88, 151 (1996)

[6] Lara-Curzio E., Watkins T., Trejo R., Luttrell C., Radovic M., Lannutti J. and England D., "Effect of Temperature and H_2-Induced Reduction on the Magnitude of Residual Stresses in YSZ-NiO/YSZ Bi-Layers", Proceedings of 106th Annual Meeting & Exposition of The American Ceramic Society (2004), in print

[8]Masashi M., Yamamoto T., Itoh H., Inaba H. and Tagawa H., "Thermal Expansion of Nickel-Zirconia Anodes in Solid-Oxide Fuel Cells during Fabrixation and Operation", J. Electrochem. Soc. 145 (1998) 1374

[9]Muller A.C., Pei B., Weber A. and Ivers-Tiffe E., "Properties of Ni/YSZ Cermet Depending on Their Microstructure", *HTMC IUPAC* July (2000)

[10]Itoh H, Yamamoto T. and Mori Masashi, "Configurational and Electrical Behavior of Ni-YSZ Cermet with Novel Microstructure for Solid Oxide Fuel Cell Anodes", J.Electrochem. Soc. 144 (1997) 641

[11]Standard Test Methods for Apparent Porosity, Water Absorption, Apparent Specific Gravity, and Bulk Density of Burned Refractory Brick and Shapes by Boiling Water", ASTM standard C20-00, (2000)

[12] Standard Test Method for Dynamic Young´s Modulus, Shear Modulus, and Poisson´s Ratio for Advanced Ceramics by Impulse Excitation of Vibration, ASTM standard C1259-01, (2001)

[13]Migliori A. and Sarro J.L., "Resonant Ultrasound Spectroscopy: Applications to Physics, Materials Measurements and Nondestructive Evaluation", John Willey & Sons, Inc. New York (1997)

[14]Szekely J. and Evans J.W., "Studies in Gas-Solid Reactions: Part I. A Structural Model for the Reaction of Porous Oxides with a Reducing Gas", Metall. Trans. **2**, 1961 (1971)

[15]Szekely J. and Evans J.W., "Studies in Gas-Solid Reactions: Part II. An Experimental Study of Nickel Oxide Reduction with Hydrogen", Metall. Trans. **2**, 1969 (1971)

[16]Szekely J. and Evans J.W., "A Structural Model for Gas-solid Reactions with Moving Boundary", Chem. Eng. Sci. **25**, 1091 (1970)

[17]Evans J.W., Song S. and Loen-Sucre C.E., "The Kinetics of Nickel Oxide Reduction by Hydrogen; Measurements in Fluidized Bed and in Gravimetric Apparatus", Metall. Trans. **7B**, 55 (1976)

[18]Sridhar S, Sichen D and Seetharaman S, " Investigation of the Kinetics of Reduction of Nickel Oxide and Nickel Aluminate by Hydrogen", Z. Metallkd. **85**, 616 (1994)

[19]Selcuk A. and Atkinson A., "Elastic Properties of Ceramic Oxides Used inSolid Oxide Fuel Cells (SOFC)", J. Euro. Ceram. Soc. **17** (1997) 1523

[20] Phani K.K. and Niygi S.K., "Young's modulus of porous brittle solids", J.Mater. Sci. **22** (1987) 257

[21]C. Liu, A. Huntz and J. Lebrun, "Origin and Development of Residual Stresses in Ni-NiO System: In-situ Studies at High Temperature by X-ray Diffraction", Matr. Sci. Eng. **A160** (1993) 113

---

* Certain commercial equipment, instruments, or materials are identified in this paper in order to specify the experimental procedure adequately. Such identification is not intended to imply recommendation or endorsement by Oak Ridge National Laboratory, nor is it intended to imply that the materials or equipment identified are necessarily the best available for the purpose.

ELECTRICAL AND MICROSTRUCTURAL INVESTIGATION OF YSZ AND TZP DOPED WITH NiO

C. L. Silva, F. R. Costa, M. R. Morelli, D. P. F. de Souza
Federal University of S. Carlos
Rod. Washington Luis Km 235
13565-905 - Sao Carlos - SP, Brazil

ABSTRACT

The NiO effect on the microstructure and electrical conductivity of 3 mol% Y_2O_3-tetragonal stabilized zirconia (TZP) and 8 mol% Y_2O_3-cubic stabilized zirconia (YSZ) was investigated. NiO doped TZP samples have smaller grain size than undoped samples after sintering up to 1500°C. At higher sintering temperature grain growth increases followed by tetragonal→monoclinic phase transformation during sintering process resulting in a large amount of cracks in the microstructure. The grain electrical conductivity of the NiO doped TZP samples was found to be independent of the NiO concentration for samples sintered up to 1500°C. The NiO doped YSZ samples maintain their cubic phase even when soaked for 8.0 h at 1600°C however, their grain and grain boundary electrical conductivity is reduced with the NiO content and the sintering temperature increase.

INTRODUCTION

Stability of the electrical and mechanical properties of solid oxide fuel cells (SOFC) components due to contacts among different materials during the fabrication and cell operation steps are a major concern.[1-4] Contacts at high temperatures may promote diffusion of ions from one component to another. Of special importance is the contact between the yttria stabilized zirconia solid electrolyte and the Ni-ZrO_2 anode cermet leading the diffusion of NiO in the zirconia grains. The main concern with the reaction between NiO and zirconia are the formation of resistive phases and modification of the structural stability of the zirconia phase. In their study of the structural stability of cubic zirconia when alloyed with NiO in the molar range of 0 to 15 %, Chen et al[5] found the presence of Ni in the ceramic grains, however, the cubic structure was maintained even for samples that have been treated at 1600°C. Kondo et al[6] studied the stability of the zirconia tetragonal phase when alloyed with NiO in the range 0 to 1.5 mol %. It was found that NiO promotes grain growth and above 0.3 mol % addition induced tetragonal to monoclinic phase transformation during the sintering process. The above results are yet not fully understood.

Taking into account that Ni-YSZ and Ni-TZP cermets are good candidates to be anode electrodes for SOFCs, the present work deals with the effect of NiO doping on the electrical conductivity and microstructure of cubic (YSZ) and tetragonal (TZP) zirconia ceramics.

EXPERIMENTAL PROCEDURE

Commercial 3 mol% yttria-stabilized tetragonal zirconia (TZ3Y) and 8 mol% yttria-stabilized cubic zirconia (TZ8Y), both Tosoh-Japan, were mixed with Nickel nitrate in isopropyl alcohol to produce 2.0, 4.0, 5.5 and 9.0 % NiO molar concentrations. The compositions and the label are shown on Table I. Each of the mixtures was heated at 800°C for 2.0 h to fully decompose the nitrate to NiO. The calcined powder was deagglomerated in isopropyl alcohol with 2.0 wt% of PVB. After drying the powders were sieved through 80 mesh. Discs were isostatically pressed at 200 MPa and sintered between 1400 and 1600°C for 1.0 h and at 1600°C for 8.0 h. The sintered discs had their densities measured by Archimedes technique and were characterized by X-ray diffraction (XRD), impedance spectroscopy (IS), and scanning electron microscopy (SEM). IS measurements were conducted in the frequency range 5 Hz to 13 MHz in the temperature range of 250-500°C. The grain size of the sintered samples was measured applying the linear intercept method on the SEM micrographs.

Table I. Compositions and respective labels

NiO content (mol%)	0.0	2.0	5.5	9.0
Label for NiO doped TZ8Y	YSZ	YSZ-2NiO	YSZ-5.5NiO	YSZ-9NiO
Label for NiO doped TZ3Y	TZP	TZP-2NiO	TZP-5.5NiO	TZP-9NiO

RESULTS AND DISCUSSION

Tetragonal Zirconia Polycrystals (TZP)

Density and crystalline phases

Discs densities for the different sintering temperatures are shown in Figure 1. All samples for which the sintering soaking time were of 1.0 h show a small decrease in density, while the samples sintered at 1600°C for 8.0 h a pronounced density decrease was observed, except the samples whose composition are free of NiO.

The XRD diffraction pattern for composition free of NiO sintered at 1600°C-8h shows only tetragonal crystalline phase, Figure 2. The XRD diffraction patterns for TZP-4NiO and TZP-9NiO samples sintered at 1600°C for two different soaking times, 1.0 and 8.0 h, are shown in Figure 3. The intensity of diffraction peaks of monoclinic phase increases with the increase of NiO concentration and the soaking time. The monoclinic phase content was calculated according Toraya[7] and is shown in Table II. The XRD pattern of samples sintered at temperatures below 1600°C show only the tetragonal phase pattern, independently of the NiO concentration.

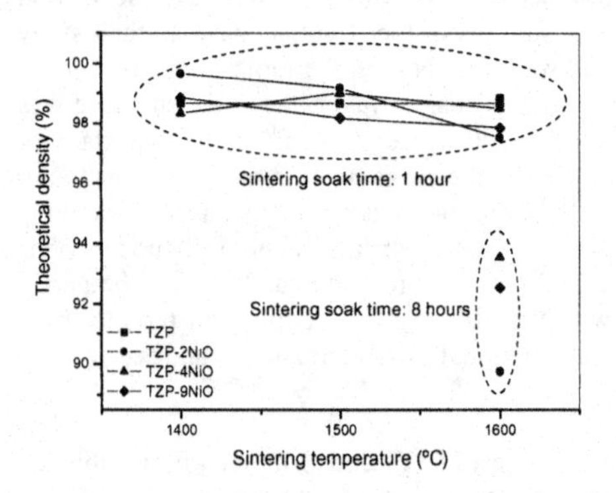

Figure 1. Body density of TZP and NiO doped compositions as a function of the sintering temperature

Figure 2. XRD pattern of TZP composition sintered at 1600°C - 8 h

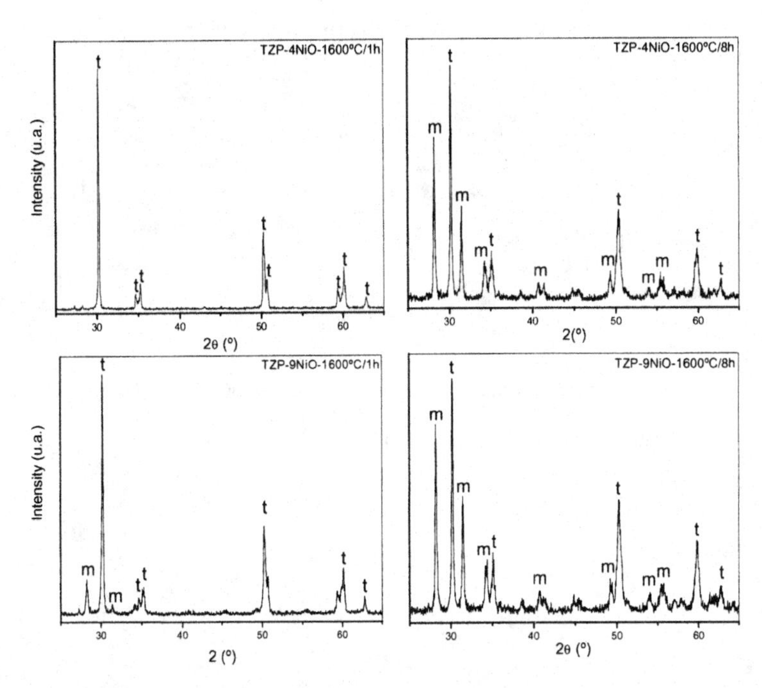

Figure 3. XRD pattern for TZP-4NiO and TZP-9NiO sintered at 1600°C during 1 and 8 h (t = tetragonal phase; m = monoclinic phase)

TableII. Percentage of monoclinic phase of the NiO doped samples

Composition	Monoclinic phase (%)	
	1 hour	8 hours
TZP-2NiO	4	62
TZP-4NiO	3	56
TZP-5.5NiO	6	53
TZP-9NiO	15	57

Microstructure of TZP samples

The SEM analysis of the polished and thermally etching surface shows that the NiO solid solution inhibited the grain growth in the samples sintered at 1400 and 1500°C while their grain size show a narrow distributions, as Figure 4 shows for the TZP-4NiO composition. Increasing the sintering temperature to 1600°C for 1.0 h a small number of very large grains are found dispersed in the microstructure while for 8.0 h of sintering time the number of large grains increases all over the sample as shows Figure 5. Figure 6 shows the microstructure of TZP-4NiO composition obtained by BSE mode where cracks are observed preferentially along the grain boundary as a result of the tetragonal to monoclinic phase transformation, Figure 3. Crack development explains the density decrease of samples sintered at 1600°C for 8.0 h, see Figure 1.

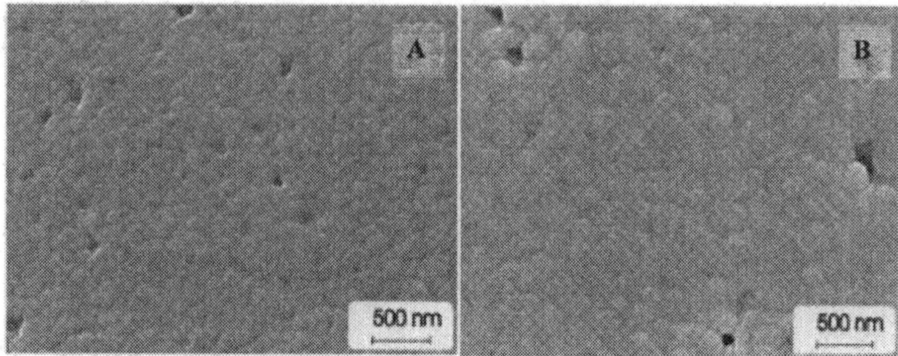

Figure 4. SEM micrographs of composition TZP-4NiO sintered during 1 h at: A) 1400°C; B) 1500°C

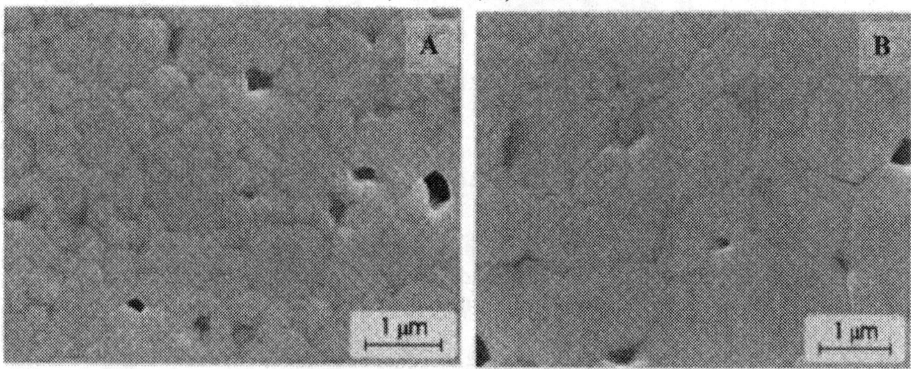

Figure 5. SEM micrographs of composition TZP-4NiO sintered at 1600°C during: A) 1 h; B) 8 h.

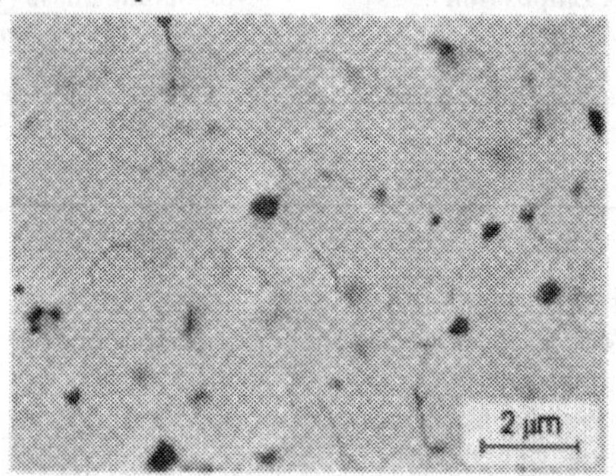

Figure 6. Backscattered electron image of composition TZP-4NiO sintered at 1600°C during 8.0 h.

All samples having NiO in their compositions have similar sintering behavior to the TZP-4NiO sample: smaller grain size than the TZP after sintering at 1400 and 1500°C; exaggerated grain growth during sintering at 1600°C followed by catastrophic destabilization of tetragonal phase, generating cracks preferentially along the grain boundary. This behavior predominates as the NiO concentration increases. Table III shows the average grain size of samples sintered at different temperatures.

Table III. Mean grain size for TZP's compositions after sintering at several temperatures

	Grain Size (μm)			
Composition	1400°C - 1h	1500°C - 1h	1600°C - 1h	1600°C - 8h
TZP	0.30	0.36	0.38	0.40
TZP-2NiO	0.10	0.14	0.29	0.73
TZP-4NiO	0.09	0.16	0.30	0.73
TZP-5.5NiO	0.09	0.15	0.33	0.73
TZP-9NiO	0.10	0.16	0.38	0.80

Electrical conductivity of TZP samples

The impedance spectroscopy spectra of the TZP samples was found to be compatible with the observed microstructures. The IS spectra of samples sintered at temperatures below 1600°C are typical of homogeneous microstructures and the semicircles of the Nyquist plot have low depression angles. On the other hand the IS spectra of samples sintered at 1600°C for 1.0 h have high depression angles due to the developed cracks. Figure 7 compares the Nyquist diagrams of compositions TZP and TZP-4NiO. The IS spectra of the others NiO compositions are similar to that shown by the TZP-4NiO sample.

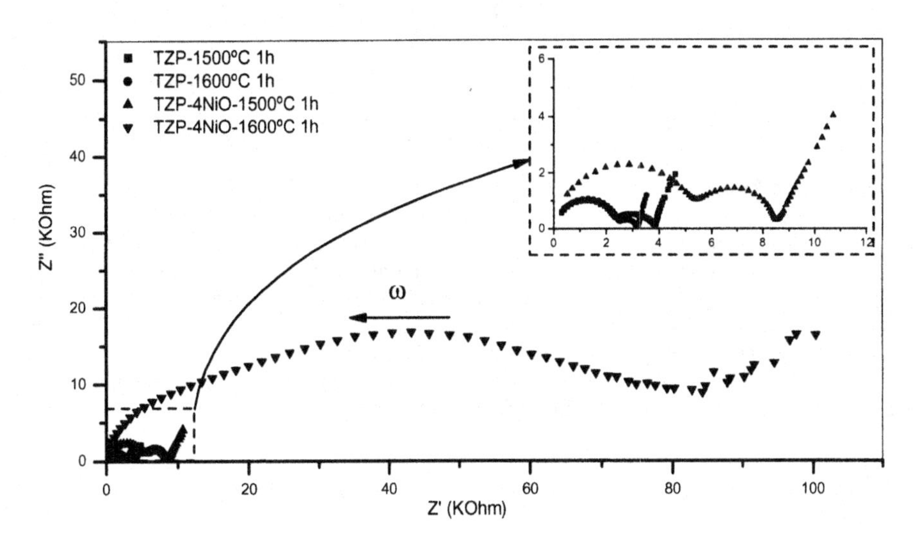

Figure 7. Nyquist plots at 400°C for compositions TZP's sintered at 1500 and 1600°C.

The grain conductivity, σ_g, of all compositions sintered at 1400 and 1500°C was calculated following the brick layer model, $\sigma_g = L / A R_g$, where L and A are the thickness and the sample contact area for the electrical measurements, respectively. R_g is the grain electrical resistance as measured directly from the Nyquist plot shown in Figure 7. The grain conductivities measured at 400°C of samples with different NiO concentration are shown in Figure 8. It can be seen that for these samples the grain conductivity is independent of the NiO concentration, indicating that concentration and mobility of the charge carriers, the oxygen vacancies, were not changed by NiO. This results points to the absence of NiO solid solutions in the zirconia lattice. Therefore, for the temperatures and sintering times used in this work, NiO must be distributed along the grain boundary inhibiting the grain growth, see Table III. The NiO solid solutions occurred only for sintering at 1600°C when it was observed destabilization of the tetragonal phase.

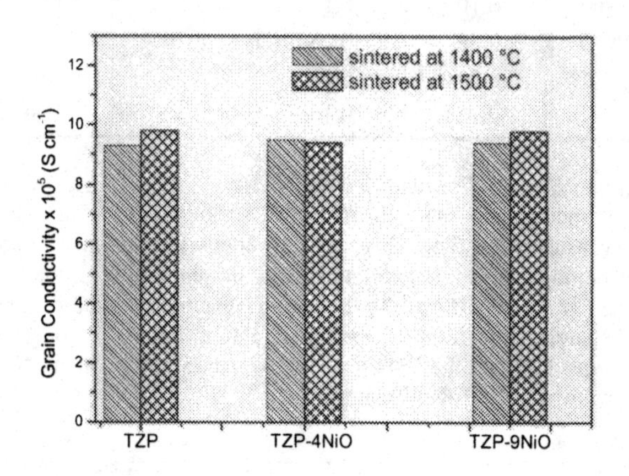

Figure 8. Grain conductivity for compositions TZP's with different NiO content

Yttria Stabilized Zirconia (YSZ)

Microstructure of YSZ samples.

The effect of NiO addition was found to be quite different for the YSZ and TZP samples. The cubic phase was found to be stable even after sintering at 1600°C. The effect observed on the microstructure was the same as previously observed by Chen et al[5]. For low NiO concentrations grain growth was observed, but, when the NiO concentration increases the microstructure becomes porous and grain size is reduced. Figure 9 shows the grain size as a function of the NiO content for samples sintered at several temperatures. The maximum grain size was detected at around 2 mol% of NiO for all sintering temperatures.

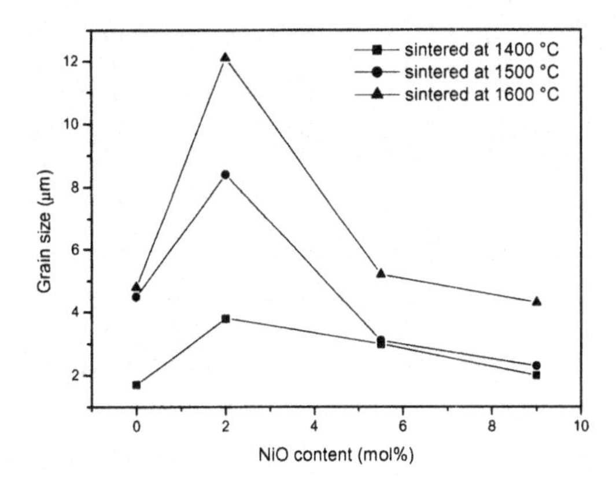

Figure 9. Grain size for compositions YSZ's sintered at several temperatures

Electrical conductivity of YSZ samples.

Large changes on the impedance spectra, IS, were not observed either by increasing the NiO concentration or the sintering temperature. However, it was observed that the contribution of grain boundary resistance to sample total resistance increased with the NiO concentration. Figure 10 shows the Nyquist plots for samples measured at 400°C for the YSZ, YSZ-2NiO, and YSZ-9NiO sintered at 1500 and 1600°C during 1 h.

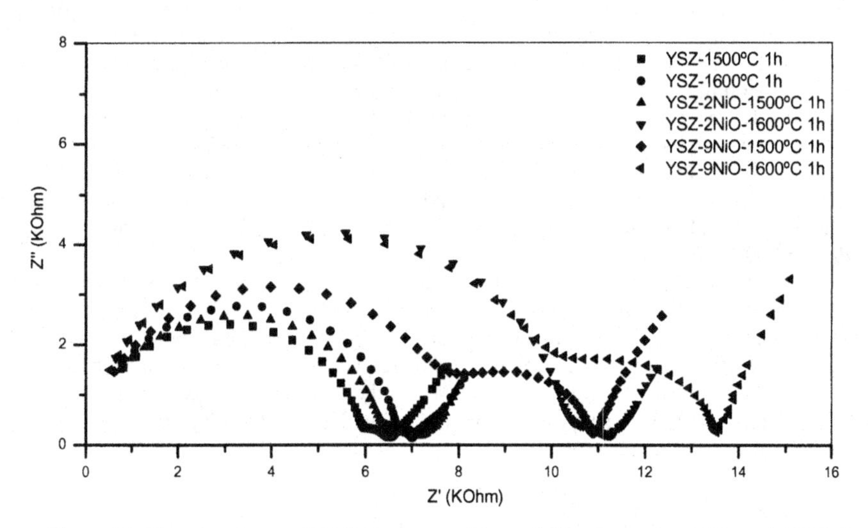

Figure 10. Nyquist plots at 400°C for compositions YSZ's sintered at 1500 and 1600°C.

The activation energies for electrical conduction of the grain and grain boundary were calculated from the Arrhenius plots. Table IV shows the values for samples sintered at three different temperatures. The YSZ-2NiO sample has the smaller activation energies for the grain and grain boundary. However, there was not found a systematic decrease in the activation energy with the increase of NiO concentration.

Table IV. Activation energy for grain and grain boundary electrical conduction for YSZ's compositions

Composition	1400°C		1500°C		1600°C	
	Ea_g (eV)	Ea_{gb}eV)	Ea_g (eV)	Ea_{gb}eV)	Ea_g (eV)	Ea_{gb}V)
YSZ	1,08	1,16	1,04	1,12	1,15	1,06
YSZ-2NiO	1,01	1,08	1,04	1,02	1,06	1,06
YSZ-5.5NiO	1.10	1.17	1.11	1.15	1.11	1.17
YSZ-9NiO	1,10	1,16	1,11	1,16	1,15	1,18

Figure 11 shows the behavior of the grain conductivity, σ_g and the effective grain boundary conductivity, σ_{gb}^{ef}, versus NiO concentration for three sintering temperatures. Both, σ_g and σ_{gb}^{ef} decrease when the NiO concentration and sintering temperature increase.

According to Chen et al[5] the zirconia lattice contracts with the dissolution of NiO, therefore, the activation energy for ionic charge carriers, as the oxygen ion vacancy, should increase. However, Table IV shows that activation energy, as measured in the present work, does not increase. On the contrary, the activation energy of the grain and grain boundary of the YSZ-2NiO sample decrease. Since the conductivity dependence on the concentration of charge carriers and mobility, the results shown in Figure 11A points to a reduction on the free oxygen ion vacancy. It seems feasible to consider that the binding energy between the oxygen vacancies and the Y^{+3} ion have increased. Because the ionic radius of the positive ions present in the lattice are different, $Y^{+3} > Zr^{+4} > Ni^{+2} > Ni^{+3}$ it may considered that Ni^{+2} and/or Ni^{+3} ions is associated with the Y^{+3} to decrease internal misfit in the lattice and provide a larger binding energy with the vacancies. Electrical conductivity measurements in annealed samples must be done to confirm the dopants association.

As shown in Figure 11B, the grain boundary conductivity decreases with the NiO content increases. This behavior suggests that the dissolution of the Ni^{+2} and/or Ni^{+3} ions in the stabilized zirconia lattice contributes for the increase of the Y^{+3} segregation to grain boundary, as it is well established in the literature[8]. This segregation increases the space charge layer according to original Frenkel's model.[9] The space charge layer on the grain boundary is rich in immobile Y^{+3} and deficient in mobile $V_O^{\bullet\bullet}$ and, consequently, the electrical conductivity of the grain boundary decreases with the Y^{+3} segregation.[10]

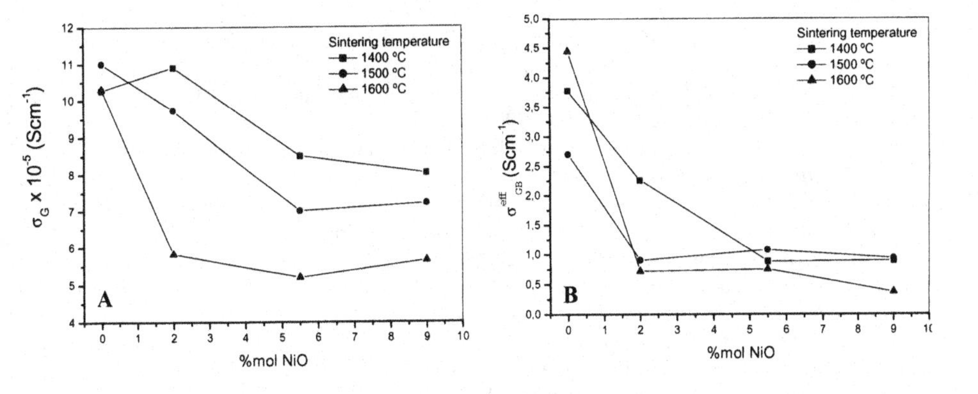

Figure 11. Electrical conductivity for YSZ's compositions sintered at several temperatures: A) grain; B) grain boundary.

CONCLUSION

The NiO doped TZP and YSZ show different behavior concerning their microstructure, XRD pattern and impedance spectroscopy spectra. From this study the following conclusions can be draw:

- NiO-TZP sintered at T< 1600°C has smaller grain size than the undoped zirconia; its grain electrical conductivity is independent of NiO content and sintering temperature.
- NiO-TZP sintered at 1600 °C show large grain size. Consequently, tetragonal → monoclinic transformation occurs cracking the sample.
- NiO-YSZ samples keep the cubic phase even for long soaking times at 1600°C and for NiO content as high as 9.0 mol%. The 2.0 mol% doped samples was found to have the larger grain size.
- The grain electrical conductivity of the NiO-YSZ decreases when the NiO content increases. As the grain conductivity activation energy is almost constant for different NiO content, the grain conductivity decrease is attributed to a decrease on the free oxygen vacancy concentration in the temperature range where the measurement have been done.
- The effective grain boundary conductivity decreases when the NiO concentration increases, what can be understood as due to further segregation of the Y^{+3} ion increasing the space charge effect that causes the reduced conductivity of the grain boundary.

ACKNOWLEDGEMENTS

To Brazilian financial agencies FAPESP (grant 00/14842-3) and CNPq (grants 477600/01-8 and 401227/2003-0)

REFERENCES

[1]A. Weber, E. Ivers-Tiffée, "Materials and Concepts for Solid Oxide Fuel Cells (SOFCs) in Stationary and Mobile Applications" *Journal of Power Sources* **127** [1-2] 273-83 (2004).

[2]T. Mori et al., " Mechanism Between Lanthanum Manganite and Yttria Doped Cubic Zirconia" *Solid State Ionics* **123** 113-119 (1999).

[3]M. B-Ricoult and M-F. Trichet "Interfacial Chemistry at Metal Electrode-Oxide Contacts" *Solid State Ionics* **150** 143-456 (2002).

[4]S. P. Simer, J. P. Shelton, M. D. Anderson, J. W. Stevenson "Interaction Between La(Sr)FeO_3 SOFC Cathode and YSZ Electrolyte", Solid State Ionics **161** 11-18 (2003).

[5]S. Chen, W. Deng, and P. Shen, "Stability of Cubic ZrO_2 (10 mol% Y_2O_3) when Alloyed with NiO, Al_2O_3 or TiO_2: Implications to Solid Electrolytes and Cermets" *Materials Science and Engineering* **B22** 247-255 (1994).

[6]H. Kondo et al, "Solid-Solution Effects of a Small Amount of Nickel Oxide Addition on Phase Stability and Mechanical Properties of Yttria-Stabilized Tetragonal Zirconia Polycrystals" *Journal of the American Ceramic Society* **86** [3] 523-25 (2003).

[7]H. Toraya, M. Yoshimura, S. Somika, "Calibration Curve for Quantative Analysis of the Monoclinic-Tetragonal System by X-Ray Diffraction" *Journal of the American Ceramic Society* **67** [6] C-119-C-121 (1984).

[8]A. E. Hughes, "Segregation in Single-Crystal Fully Stabilized Yttria-Zirconia", *Journal of the American Ceramic Society* **78** [2] 369-78 (1995).

[9]J. Frenkel, *Kinetic Theory of Liquids*, Oxford University Press, Oxford, 1946

[10]J-C. M'Peko and M. F. de Souza, "Ionic Transport in Polycrystalline Zirconia and Frenkel's Space-Charge layer Postulation", *Applied Physics Letters* **83** [4] 737-739 (2003).

CHEMICAL SYNTHESIS OF LSGM POWDERS FOR SOLID OXIDE FUEL CELL (SOFC) ELECTROLYTE

Cinar Oncel and Mehmet Ali Gulgun
Sabanci University, FENS, Tuzla, 34956
Istanbul, Turkey

ABSTRACT

Synthesis of LSGM ($La_{0.9}Sr_{0.1}Ga_{0.8}Mg_{0.2}O_{3-\delta}$), LSFM ($La_{0.9}Sr_{0.1}Fe_{0.8}Mg_{0.2}O_{3-\delta}$), and LSCM ($La_{0.9}Sr_{0.1}Cr_{0.8}Mg_{0.2}O_{3-\delta}$) powders were achieved via organic precursor method. Different organic "carrier" molecules were used for powder synthesis. Citric acid, tartaric acid, Pechini precursors, polyvinyl alcohol, and ethylene diaminetetraacetic acid were selected as organic carriers for their ability to stabilize the metal ions. Each organic carrier material exhibited a different degree of effectiveness in the synthesis of the mixed oxide powders. One of the main factors affecting the phase purity appears to be the interaction of the functional groups with the constituent cations. The effectiveness of the organic carrier with varying number and type of functional groups is evaluated and discussed in terms of the phase distribution in the powders after the calcination step.

INTRODUCTION

Solid oxide fuel cells are regarded as the energy production systems for 21th century due to their high efficiency, utilization of a variety of the fuel resources, and environmental friendliness. Strontium and magnesium-doped lanthanum gallate (LSGM, e.g. $La_{1-x}Sr_xGa_{1-y}Mg_yO_{3-(x+y)/2}$) is a perovskite-type oxide and one of the most promising electrolyte materials for Intermediate Temperature-SOFC applications. Its ionic conductivity values are much higher than the one of YSZ electrolyte, and comparable to that of ceria-based electrolytes in the high and intermediate temperature ranges. Ionic conductivities of YSZ, LSGM, and CGO electrolytes at temperatures 600^{0}C, 800^{0}C, and 1000^{0}C are tabulated in Table 1.

Table 1. Ionic conductivities of YSZ, LSGM, and CGO for 600^{0}C, 800^{0}C, and 1000^{0}C.

Electrolyte	600^{0}C	800^{0}C	1000^{0}C
YSZ	0.003 S/cm[(1)]	0.03 S/cm[(1)]	0.1 S/cm[(1)]
LSGM	0.02 S/cm[(2)]	0.12 - 0.17 S/cm[(1)]	0.25 S/cm[(2)]
CGO	0.025 S/cm[(1)]	0.1 S/cm[(2)]	0.25 S/cm[(1)]

Each electrolyte material in SOFC construction is designed to exhibit the best performance under SOFC operating conditions. Small discrepancies in the composition results in a poorer performance of SOFC. For example, excellent ionic conductivity was achieved with Sr and Mg-doped $LaGaO_3$ electrolyte material of the following composition ($La_{0.9}Sr_{0.1}Ga_{0.83}Mg_{0.17}O_{3-\delta}$)[3]. Small deviations from composition resulted in a decrease in the ionic conductivity. Therefore, it is important to produce pure and single-phase SOFC components with the desired compositions.

The organic precursor technique is a method widely used in mixed oxide powder synthesis[4-7]. The predicted mechanism in organic precursor method for achieving a stable precursor is the chelating/complexing of the metal cations by the functional groups of the organic carrier materials in the solution[8]. This stabilization action is believed to be due to columbic attraction forces between the carboxylic or hydroxyl groups of the carrier materials and metal cations. The molecular geometry of the functional groups is also believed to play an important role in the chelating/complexing ability. As a result of this stabilization, metal cations are distributed homogeneously in the solution and are stabilized in the pre-ceramic precursor after solvent removal. During calcination, after the organic burn-out, an amorphous powder is obtained. At higher temperatures, crystallization of the desired phases takes place. Due to the homogeneity in molecular level, lower diffusion distances for the cations are required to obtain the desired crystal phase. This in turn may result in lower temperatures for phase purity compared to the solid-state reaction technique[9]. Moreover, combustion of the organic materials results in local temperature increases, that help diffusion process and final crystallization.

Different types of the organic carrier materials can be used in oxide synthesis. One of the successful techniques for single phase mixed oxide powders is Pechini process[10]. Pechini process operates through polyesterfication between hydrocarboxylic acids such as citric acid, and polyhydroxy alcohols such as ethylene glycol[10]. According to the ester reaction shown below, carboxyl end of citric acid and hydroxyl end of ethylene glycol react and a water molecule is released. The acid acts as a chelating agent that stabilizes the cations dissolved in the solution.

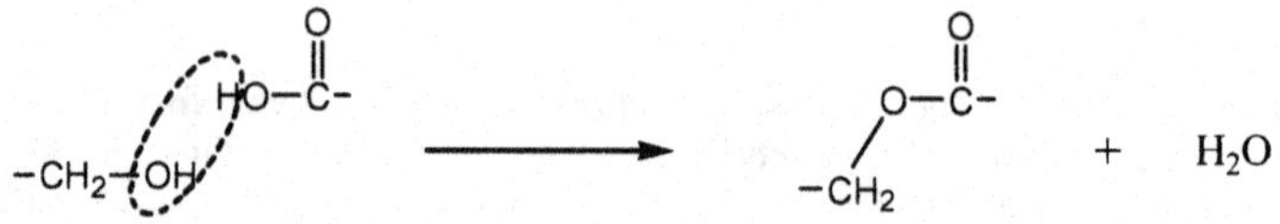

$$-CH_2-OH + HO-C(=O)- \longrightarrow -CH_2-O-C(=O)- + H_2O$$

Figure 1. Ester reaction.

The polymerization is based on the polyesterfication between the metal-chelate complexes and polyhdroxyl alcohols. By the polyesterification process, randomly coiled macromolecular chains are obtained. These chains may chelate cations uniformly and form very stable metal-organic complexes. Moreover, due to chelating action and high viscosity polymeric network, cation segregation during solvent evaporation is hindered. The resultant ceramic powders possess better chemical homogeneity and smaller particle size. Organic precursor methods with different organic carrier materials, such as citrate synthesis, polymeric precursor synthesis, and urea method have been employed in various oxide syntheses. However there are few studies on the chemical synthesis of SOFC components via different carrier materials[11,12]. In this study LSGM synthesis was conducted via the organic precursor method. Also the effects of different carrier materials on phase purity and crystallization behavior were investigated. Additionally, $La_{0.9}Sr_{0.1}Fe_{0.8}Mg_{0.2}O_{2.85}$ (LSFM) and $La_{0.9}Sr_{0.1}Cr_{0.8}Mg_{0.2}O_{2.85}$ (LSCM) powders were synthesized with the same synthesis route and organic carrier materials. In the synthesis of LSFM and LSCM powders, the stoichiometry used in LSGM synthesis was kept to investigate the effects of different cations (Fe^{3+} or Cr^{3+} in place of Ga^{3+}). Iron and chromium were chosen to replace gallium such that the new materials can be evaluated as candidates for SOFC interconnect and cathode materials.

EXPERIMENTAL PROCEDURE

Cation sources were nitrate salts of the desired cations selected for their high solubility in cold water. Lanthanum nitrate hexahydrate ($La(NO_3)_3 . 6H_2O$, >99%, Sigma Aldrich Chemie GmbH, Taufkirchen, Germany), gallium nitrate nanohydrate ($Ga(NO_3)_3 \cdot 9H_2O$, 99.9%, ChemPur Feinchemikalien und Forschungbedarf GmbH, Karlsruhe, Germany), strontium nitrate ($Sr(NO_3)_2$, >99%, Sigma Aldrich Chemie GmbH, Taufkirchen, Germany), magnesium nitrate hexahydrate ($Mg(NO_3)_2 \cdot 6H_2O$), iron nitrate nanohydrate ($Fe(NO_3)_3 \cdot 9H_2O$), chromium nitrate nanohydrate ($Cr(NO_3)_3 \cdot 9H_2O$, all three salts>99%, Merck KgaA, Darmstadt, Germany) were the sources of lanthanum, gallium, strontium, magnesium, iron, and chromium, respectively.

Polyvinyl alcohol, PVA ($n \cdot C_2H_4O$, MW = 72000, >98%, Merck KgaA, Darmstadt, Germany), citric acid ($C_6H_8O_7$, >99.5%, Sigma Aldrich Chemie GmbH, Taufkirchen, Germany), ethylenediaminetetraacetic acid, EDTA ($C_{10}H_{16}N_2O_8$, 98%, Sigma Aldrich Chemie GmbH, Taufkirchen, Germany), tartaric acid ($C_4H_6O_2$, 99.7%, Sigma Aldrich Chemie GmbH, Taufkirchen, Germany) and ethylene glycol, EG ($C_2H_6O_2$, 99.5%, Carlo Erba Reagenti, Mendetison Group) were the organic/polymeric materials used as "carriers" for the cations. Distilled water and nitric acid (HNO_3, 65% solution, Sigma Aldrich Chemie GmbH, Taufkirchen, Germany) were utilized as solvents in the experiments where indicated.

Low temperature chemical synthesis of three different mixed oxides with four-cations, $La_{0.9}Sr_{0.1}Ga_{0.8}Mg_{0.2}O_{2.85}$ (LSGM), $La_{0.9}Sr_{0.1}Cr_{0.8}Mg_{0.2}O_{2.85}$ (LSCM), and $La_{0.9}Sr_{0.1}Fe_{0.8}Mg_{0.2}O_{2.85}$ (LSFM) was investigated to produce these mixed-oxides as single phase, fine powders. For the synthesis, desired amounts of cation salts were dissolved in distilled water/organic carrier solutions to obtain exact stoichiometry. In the organic precursor route with organic/polymeric carrier materials, the amounts of carrier materials were determined to obtain 1:1 cation to organic molecule ratio in the solution. For the synthesis of LSGM, two different synthesis concepts were applied. In both routes, nitrate salts of each constituent cation were selected as the cation source. Polyvinyl alcohol (PVA, Steric entrapment method[13]), citric acid (CA), tartaric acid (TA), ethylene diaminetetraacetic acid (EDTA), or Pechini precursors (with 60% citric acid – 40% and ethylene glycol, or 90% citric acid – 10% ethylene glycol mixtures) were used as the organic carrier materials in solution. In the synthesis of LSGM with EDTA as the carrier material, nitric acid was the solvent. In the second production route, the nitrate salt of each constituent cation was dissolved in distilled water without any organic molecule. Solutions were mixed with a magnetic stirrer and heated up to 300°C to evaporate the solvents and to obtain a crisp powder. These organo-metallic precursors were ground and calcined in pure alumina crucibles at 700°C, 800°C, 900°C, 1000°C, 1050°C, 1100°C, 1150°C, or 1200°C in a box furnace (in air) with 10°C/min heating rate. After reaching the final temperature the furnace was turned of immediately and the powders were allowed to cool in the furnace.

In LSFM and LSCM synthesis, calcination temperatures, 500°C, 550°C, 600°C, 650°C, 700°C, 750°C, 800°C, and 850°C were used. Crystal structure and phase distribution of the powders at room temperature were studied with an x-ray powder diffractometer (Bruker AXS-D8, Karlsruhe, Germany). The measurements were performed in the 2θ range of 10° - 90° at 40 kV and 40 mA, using Cu-K_α radiation. In all measurements, the step size was 0.03°, and data collection period was 2 seconds in each step. $K\alpha_2$ peaks were suppressed in the x-ray diffraction measurements. In x-ray diffraction plots, the percentages of each phase were calculated by taking the ratio of the height in the intensity axis of the main peak (100% peak) of each phase, to the sum of the height of the main peaks of all phases. Prior to the peak-height measurements, background subtraction was performed. For phase identification, the experimental spectra were compared to the characteristic x-ray card files in the JCPDS database. For phases that were synthesized in this work for the first time, the experimental spectra were compared to the JCPDS file for the compound from which the new mixed cation oxide was derived, i.e. the LSFM x-ray spectrum was compared to the JCPDS file for $LaFeO_3$ and $La_{0.9}Sr_{0.1}FeO_{2.95}$; while the LSCM x-ray spectrum was compared to the JCPDS file for $LaCrO_3$.

RESULTS and DISCUSSION

LSGM Synthesis

Tables 2-4 list the percent amounts of phases in the powders obtained with a different organic carrier molecule and calcined at the specified temperatures. The amounts are determined from the XRD intensities. Figure 2 illustrates the effectiveness of the carrier molecules in terms of LSGM phase amount in the powders as a function of calcination temperature.

Table 2. Phase percentages of powders for two different Pechini precursors.

Pechini (60-40)	*LSGM*	*$LaSrGaO_4$*	*$SrLaGa_3O_7$*	*$La_4Ga_2O_9$*	*Pechini (90-10)*	*LSGM*	*$LaSrGaO_4$*	*$SrLaGa_3O_7$*	*$La_4Ga_2O_9$*
800°C	0,0%	0,0%	60,0%	40,0%	**800°C**	5,0%	3,4%	62,8%	28,8%
900°C	12,8%	0,6%	39,2%	47,4%	**900°C**	76,5%	4,5%	7,5%	11,5%
1000°C	53,0%	1,2%	19,7%	26,0%	**1000°C**	89,4%	2,8%	3,4%	4,4%
1050°C	67.4%	1,3%	11,2%	20,1%	**1050°C**	92,8%	5,2%	1,1%	0,9%
1100°C	85.0%	1.8%	4,4%	8,7%	**1100°C**	89,9%	7,4%	2,2%	0,5%
1150°C	95.8%	4,2%	0,0%	0,0%	**1150°C**	95,2%	4,1%	0,7%	0,0%
1200°C	94.9%	5,1%	0,0%	0,0%	**1200°C**	92,2%	6,6%	1,2%	0,0%

As the ethylene glycol amount in the Pechini precursor was decreased to 90:10 CA:EG, the amount of LSGM appeared to increase at a given calcination temperature (see Table 2). This may be due to an increased number of active carboxylic ends of citric acid compared to the 60:40 CA:EG precursor. To this end, the use of citric acid alone could give better phase distribution due to a more effective chelating action. When citric acid alone was used as the organic carrier, a strong increase in the percentage of LSGM phase was observed in powders calcined between 800°C and 900°C (Table 3). The better LSGM yield of the citric acid as the carrier material, compared to Pechini process may be explained by the large number of free active carboxylic groups of the citric acid, without the ethylene glycol to esterify with. The effectiveness of the citric acid in stabilizing cations stems from the "claw" shape arrangement of carboxylic groups of the citric acid, which is suitable to chelate cations. In Pechini process, ethylene glycol molecules attach to the carboxylic ends of citric acid molecules according to the reaction in Figure 1 and thereby decrease the number of active carboxylic ends. This in turn decreased the effectiveness of citric acid molecules for chelating La^{+3}, Sr^{+2}, Ga^{+3}, and Mg^{+2} ions in the Pechini solution.

Table 3. Phase percentages of powders for precursors wit CA and TA.

CA	*LSGM*	*$LaSrGaO_4$*	*$SrLaGa_3O_7$*	*$La_4Ga_2O_9$*	*TA*	*LSGM*	*$LaSrGaO_4$*	*$SrLaGa_3O_7$*	*$La_4Ga_2O_9$*
800°C	10,1%	1,9%	61,4%	26,7%	**800°C**	27,1%	0,0%	55,5%	17,5%
900°C	94,2%	0,8%	2,3%	2,8%	**900°C**	28,8%	0,0%	39,3%	31,8%
1000°C	96,4%	1,6%	0,5%	1,5%	**1000°C**	62,7%	0,0%	16,5%	20,8%
1050°C	96,0%	4,0%	0,0%	0,0%	**1050°C**	96,4%	0,5%	1,6%	1,5%
1100°C	92,8%	7,2%	0,0%	0,0%	**1100°C**	98.7%	1,3%	0,0%	0,0%
1150°C	*95,7%*	*4,3%*	*0,0%*	*0,0%*	**1150°C**	*99,0%*	*1,0%*	*0,0%*	*0,0%*
1200°C	95,0%	5,0%	0,0%	0,0%	**1200°C**	97,9%	2,1%	0.0%	0.0%

To see the effect of the number of carboxylic ends in the chelating cations, different acids with two or four carboxylic ends (tartaric acid and EDTA) were used as the carrier material. Tartaric acid has two carboxylic groups in its molecular structure. As mentioned before, the chelating ability of citric acid is due to its three carboxylic ends. Therefore tartaric acid with two carboxylic groups may be less effective, resulting in a lower LSGM phase amount for the same temperatures. Also the geometry of these molecules may play an important role in the chelating action. Experiments with tartaric acid confirmed the expected relationship between the number of carboxylic ends, geometry of the carrier molecule and the chelating ability. In LSGM powder synthesis with tartaric acid as the carrier material, lower amounts of LSGM phase were obtained (28.8% and 62.7% at 900^0C and 1000^0C, respectively) in comparison to the LSGM powder synthesized with citric acid (94.2% and 96.4% at 900^0C and 1000^0C, respectively) as the carrier material at low temperatures (see Table 3). However, the amount of LSGM phase calcined from TA-acid precursors at temperatures above 1050^0C was larger than those prepared with citric acid based precursors.

Table 4. Phase percentages of powders for precursors with EDTA and PVA.

EDTA	*LSGM*	*$LaSrGaO_4$*	*$SrLaGa_3O_7$*	*$La_4Ga_2O_9$*	*PVA*	*LSGM*	*$LaSrGaO_4$*	*$SrLaGa_3O_7$*	*$La_4Ga_2O_9$*
800°C	6,6%	0,0%	56,4%	37,0%	**800°C**	0,0%	0,0%	33,8%	66,2%
900°C	42,5%	3,7%	29,7%	24,1%	**900°C**	0,0%	0,0%	38,2%	61,8%
1000°C	87,2%	2,5%	5,1%	5,2%	**1000°C**	4,3%	0,0%	39,5%	56,2%
1050°C	84,4%	4,5%	7,1%	4,0%	**1050°C**	35,2%	1,5%	26,4%	36.9%
1100°C	88,2%	3,5%	5,0%	3,3%	**1100°C**	36,7%	0,6%	25,6%	37,1%
1150°C	91,9%	3,6%	3,0%	1,5%	**1150°C**	85,8%	2,7%	4,4%	7.0%
1200°C	*94.6%*	*3,2%*	*2,2%*	*0,0%*	**1200°C**	*96,5%*	*3,5%*	*0,0%*	*0,0%*

EDTA was also used as the carrier material in the LSGM synthesis. EDTA appears to be more efficient than TA but still less effective than CA (see Tables 3 and 4). If a simple relationship between the number of carboxylic acid ends of a molecule and its effectiveness in chelating various cations were to be expected, EDTA should have been a more effective carrier molecule for mixed cations.

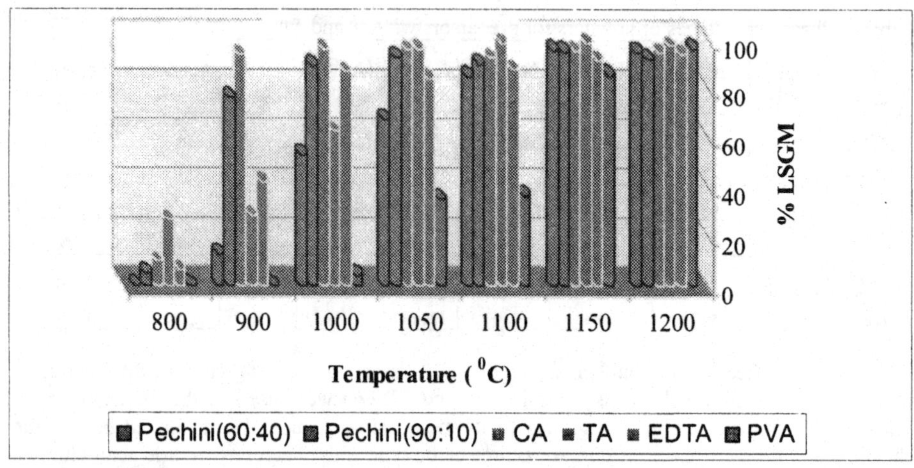

Figure 2. LSGM percentages of the powders calcined with different carriers at different temperatures.

In the experiments conducted with the carrier materials that have carboxylic ends, such as Pechini precursors, CA, TA, and EDTA; the percentage of $LaSrGa_3O_7$ phase was higher than $La_4Ga_2O_9$ phase below 1000^0C (see Tables 2-4). However, in LSGM synthesis with PVA, the percentage of $La_4Ga_2O_9$ phase was higher than the $LaSrGa_3O_7$ phase (see Table 4). $La_4Ga_2O_9$ phase is lanthanum-rich, $LaSrGa_3O_7$ phase is gallium-rich. Therefore, it appears that an interaction (complexing) of PVA with lanthanum ions may be less favored compared to gallium ions in a system composed of these four cations. Thus, an early crystallization of $La_4Ga_2O_9$ phase is more favorable than the other phases because of the loosely held La^{3+} ions in PVA-cation polymeric network at low temperatures (below 800^0C). Also, LSGM percentage in the resulting powders calcined up to 1100^0C was considerably small compared to the LSGM percentage in the powders synthesized with Pechini precursors, CA, TA, or EDTA (see Tables 2-4). LSGM powders synthesized using nitrate sources of all cations without using any complexing or chelating agent yielded much lower amounts of LSGM at any calcinations temperature. The results once more confirmed the effectiveness of carrier material in mixed oxide synthesis. The highest amount of LSGM phase obtained without any organic carrier in the process was 58.6% at 1200^0C.

LSFM and LSCM Synthesis

Since ionic radii and valences of iron and chromium ions are similar to gallium ion, substitution of these cations with gallium ion may result in powders with similar crystal structure and properties to LSGM. Moreover, $LaFeO_3$ and $LaCrO_3$ based oxides are candidate cathode and anode materials for SOFC, respectively. X-ray diffraction results of the calcined powders at different temperatures showed that, single phase LSFM powders were obtained at 550^0C with PVA, CA, or EDTA as the carrier materials. This result indicated that organic precursor method is an efficient technique to synthesize multi-cation oxides. According to the author's best knowledge, Sr-doped $LaFeO_3$, a three-cation oxide material, could be synthesized at 1200^0C with some amount of undesired phases[14]. However, synthesis of LSFM is reported in our work for the first time. In LSCM synthesis, 96.9% LSCM powders were obtained at 850^0C with PVA as the organic carrier material. Sauvet et al. synthesized Sr-doped $LaCrO_3$ at 1000^0C as single phase powder[15]. The result in this study may still seem to be promising in light of the difficulty of synthesizing a four-cation oxide rather than a three-cation oxide.

The most intriguing result of this study is the inconsistency of the effectiveness of the organic carrier materials for different powders. In LSGM synthesis CA seems to be the best and PVA seems to be the worst carrier material for low temperature synthesis. A similar tendency was also observed in LSFM synthesis. However, in LSCM synthesis, PVA was the most effective organic carrier among all the others. Moreover, CA was the one of the worst carrier material. These results emphasize the need for studying the effectiveness of "chelating" action to obtain single phase oxide powders. If a strong chelating was the only necessary criterium for desired phase formation, CA should be also effective in LSCM synthesis.

Functional group – ion (or ion group) interaction appears to be also important in desired phase formation. As underlined before, interactions of PVA with lanthanum and gallium ions were different than the other carrier materials. This is an indication of the effect of functional group – ion interaction for different carrier materials. In order to understand cation functional group interaction, Fourier Transform Infrared Spectrometry (FTIR) and Nuclear Magnetic Resonance (NMR) Spectroscopy studies will be conducted on different organo-metallic precursors. FTIR may be useful to determine compositions and relative amounts of functional groups at different calcination temperatures. NMR spectroscopy may be useful to determine chemical structure of the powders at each calcination temperature.

CONCLUSION

In this study, synthesis of LSGM, LSFM, and LSCM powders were performed via organic precursor method by using different organic carrier materials. When calcined at a low temperature ($< 1000^0$C), precursors synthesized using citric acid (CA) as the organic carrier material yielded 94,2% LSGM phase in the powders. Maximum LSGM concentration (99%) in the synthesized powders was obtained at 1150^0C using tartaric acid (TA) as the organic carrier material. In contrast to LSGM, single-phase LSFM was obtained with relative ease from the precursors calcined at 550^0C. CA appeared to be most effective precursor for low temperature synthesis of LSFM. The best concentration of LSCM phase in the synthesized powders was 96.9%, when polyvinyl alcohol (PVA) was used as the organic carrier material.

Every organic carrier material has exhibited a different performance for the synthesis of mixed oxide powders. Moreover, the performance of one organic carrier material varied for each type of mixed oxide powder synthesis. TA was the best organic carrier material for LSGM synthesis at calcination temperatures larger than 1000^0C, but it performed poorly in LSCM synthesis when compared to PVA at all calcination temperatures. Cation chelating and/or stabilizing ability of the functional groups of the organic carrier materials does not appear to be scaling just with the number of functional groups of the carrier molecule. A more complex interaction of the organic carrier with different cations may play an important role in synthesis of single phase mixed oxide powders at relatively low temperatures.

REFERENCES

[1] J. Fleig, K.D. Kreuer, J. Maier, "Handbook of Advanced Ceramics", *Materials, Applications, and Processing*, Academic Press, 1-60 (2001).

[2] Helmut Ullmann, Nikolai Trofimenko, "Composition, structure and transport properties of perovskite-type oxides", *Solid State Ionics* **119**, 1-8 (1999).

[3] S.P.S. Badwal, "Stability of solid oxide fuel cell components", *Solid State Ionics* **143**, 39-46 (2001).

[4] G.Ch. Kostogloudis, Ch. Ftikos, A. Ahmad-Khanlou, A. Naoumidis, D. Stöver, "Chemical compatibility of alternative perovskite oxide SOFC cathodes with doped lanthanum gallate solid electrolyte", *Solid State Ionics* **134**, 127-138 (2000).

[5] A. Cüneyt Taş, Peter J. Majewski, Fritz Aldinger, "Chemical preparation of pure and strontium- and/or magnesium-doped lanthanum gallate powders", *J. Am. Ceram. Soc.* **83**[12], 2954-2960 (2000).

[6] F. Riza, Ch. Ftikos, F. Tietz, W. Fischer, "Preparation and characterization of $Ln_{0.8}Sr_{0.2}Fe_{0.8}Co_{0.2}O_{3-\delta}$ (Ln=La, Pr, Nd, Sm, Eu, Gd)", *Journal of the European Ceramic Society* **21**, 1769-1773 (2001).

[7] Marko Hrovat, Ariane Ahmad-Khanlou, Zoran Samadzija, Janez Holc, "Interactions between lanthanum gallate based solid electrolyte and ceria", *Materials Research Bulletin*, Vol. **34**, Nos. 12/13, 2027-2034 (1999).

[8] Mehmet A. Gülgün, My H. Nguyen, Waltraud M. Kriven, "Polymerized organic-inorganic synthesis of mixed oxides", *J. Am. Ceram. Soc.* **82**[3], 556-560 (1999).

[9] Mehmet A. Gülgün, Oludele O. Popoola, Waltraud M. Kriven, "Chemical synthesis and characterization of calcium aluminate powders", *J. Am. Cream. Soc.* **77**[2], 531-539 (1994).

[10] M.P. Pechini, U.S. Patent No.3, 330, 697, July (1967).

[11] M. Marinsek, K. Zupan, J. Maeek, "Ni-YSZ cermet anodes prepared by citrate/nitrate combustion synthesis", *J. Power Sources* **106**, 178-188 (2002).

[12] K. Zupan, S. Pejovnik, J. Maeek, "Synthesis of nanometer crystalline lanthanum chromite powders by the citrate-nitrate autoignition reaction", *Acta Chim. Slov.* **48**(1), 137-145 (2001).

[13] Mehmet.A. Gülgün, W.M. Kriven, "Polymerized Organic-Inorganic Complex Route for Mixed-Oxide Synthesis", US patent# 6,482,387.

[14] S.P. Simner, J.R. Bonnett, N.L. Canfield, K.D. Meinhardt, V.L. Sprenkle, J.W. Stevenson, "Development of lanthanum ferrite SOFC cathodes", *Journal of Power Sources* **113**(1), 1-10 (2003).

[15] A.-L. Sauvet, J. Fouletier, F. Gaillard, M. Primet, "Doped lanthanum chromites as SOFC anode materials", *J. of Catalysis* **209**(1), 25-34 (2002).

LONG-TERM THERMAL CYCLING OF PHLOGOPITE MICA-BASED COMPRESSIVE SEALS FOR SOLID OXIDE FUEL CELLS

Yeong-Shyung Chou and Jeffry W. Stevenson
K2-44, Materials Department
Pacific Northwest National Laboratory
P O. Box 999,
Richland, WA 99352

ABSTRACT

Reliable sealants are one of the toughest challenges in advancing the solid oxide fuel cell technologies. One of the most stringent requirements for sealants is the thermal cycle stability. The sealants have to survive multiple thermal cycles during operation in stationary and transportation applications. Recently, researchers at the Pacific Northwest National Laboratory have developed a hybrid mica-based compressive seal with which leak rates were reduced to 2-4x10^{-2} to 10^{-3} sccm/cm at 800C. Long-term thermal cycling (up to 700 thermal cycles) will be conducted on the Phlogopite mica-based compressive seals. Open circuit voltage will be measured on a medium-sized (2"x2") 8YSZ plate with the mica seals during thermal cycling in a dual environment (2.75% H_2/Ar vs. air). The measured OCVs will be compared to the Nernst voltages to assess the sealing capability of the hybrid Phlogopite mica seals.

INTRODUCTION

The sealants for planar solid oxide fuel cells (SOFC) have to satisfy many stringent requirements. The sealant has to provide hermetic seal or of very low leak rates to avoid the direct mixing of the fuels and the oxidants. It has to be robust to stand the transient stresses due to rapid startup or shutdown, and the residual stresses due to mismatch in thermal expansion of different SOFC components when rigidly bonded together. It has to be thermally and chemically stable in the harsh SOFC environments, i.e., oxidizing, humid, and reducing at elevated temperatures, for long time (e.g., > 40,000 hours). It has be non-conducting and causing no degradation or corrosion to the mating materials. And most of all, it has to survive several hundreds to thousands thermal cycles during life time service in stationary or transportation applications. Currently, there are three types of approaches for SOFC seal development: rigid glass (or glass-ceramics and glass composite) seals [1-3], metallic brazes [4-5], and compressive seals [6-10]. The use of compressive seals offers the unique advantage over the other approaches in that a stringent matching of the coefficient of thermal expansion (CTE) of the various SOFC stack components is not necessary. As the thermal cycle stability appears to be the most stringent requirement for SOFC seals, it is the objective of this paper to present a long-term (up to 700 cycles) thermal cycle stability of the compressive mica seals. Recently, Pacific Northwest National Laboratory has developed a novel "hybrid" mica compressive seals that the high temperature leak rates can be greatly reduced as compared to the as-received plain micas [6]. In this paper we will present the first study of long-term thermal cycle stability of three "hybrid" micas. Open circuit voltage (OCV) will be used to characterize the thermal cycle stability of the hybrid micas used on a standard (2"x2") 8YSZ electrolyte plate. In addition, leak rates will be experimentally measured and compared to the estimations from OCVs.

EXPERIMENTAL

Materials

The mica used in this study is a commercially available Phlogopite mica paper (McMaster-Carr, Atlanta, GA). The mica paper is composed of discrete mica flakes overlapping with each other. Two Phlogopite mica papers were used in this study. One is about 0.004" thick and contains 3~5% of organic binders (mica-A). The other is about 0.003" thick and contains no binders (mica-B). Both materials have similar color and surface textures. In this paper, three hybrid micas, i.e., the mica was sandwiched between two glass layers [6], were tested for the long-term thermal cycle stability. Sample #1 and #2 used mica-A in the hybrid form but were pressed at different stresses, i.e., 100 psi for sample #1, and 50 psi for sample #2. For the hybrid micas, the glass layers were made by tape casting of a Ba-Al silicate glass which had a CTE closely matched with the anode-supported electrolyte bilayers. In our earlier study [6], the extra glass layers in the hybrid mica would greatly reduce the leak rate as compared to the as-received micas. In the third test of long-term thermal cycling (sample #3), we used a glass-mica composite. The glass-mica composite was fabricated with a composition of 20v% of a Ba-Al silicate glass and 80v% of the Phlogopite mica (mica-B). The glass-mica composite was made by mixing appropriate amount of mica flakes with the attrition-milled glass powders in water. A small amount (2 wt%) of aqueous binder was added in the slurry. The slurry was poured onto a plastic mold with Mylar film lining. The slurry was dried at room temperature for a couple of hours before cutting into desired shape and size for thermal cycling and OCV tests.

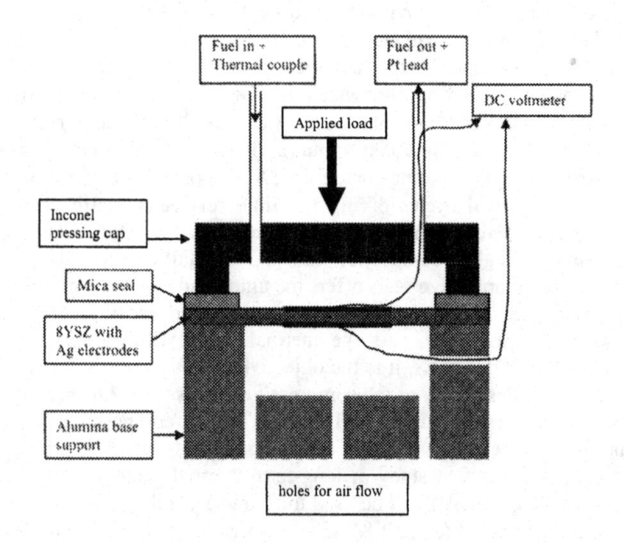

Fig. 1. Schematic showing the test fixture for open circuit voltage test of dense 8YSZ plate with compressive mica seals.

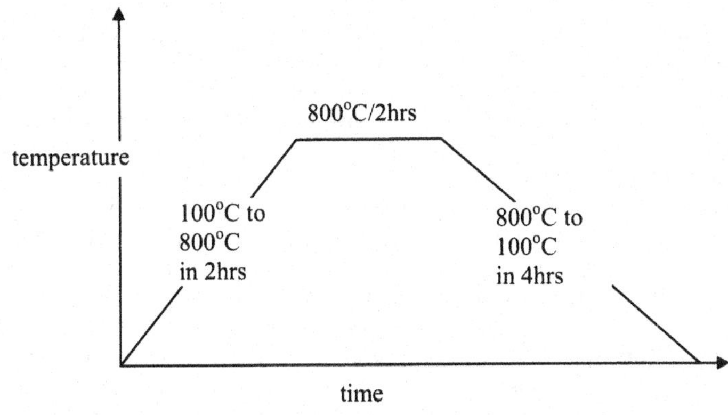

Fig. 2. Typical temperature profile for thermal cycling of sample #2, and sample #3. Sample #1 was heated more rapidly (100°C to 800°C in 35 minutes) and cooled slowly (in 5 hrs and 25 minutes to 100°C).

Open Circuit Voltage Test and Thermal Cycling

In order to assess the sealing capability of the mica seals, open circuit voltage (OCV) tests were conducted on in-house made 8YSZ electrolyte plates with the three hybrid micas. 2"x2" dense 8YSZ plates were prepared by slip casting fine 8YSZ powders (TOSOH, Zirconia, TZ-8Y, Japan), followed by sintering at 1450°C for 2 hours. The sintered plates were machined to the desired size (2"x2") and thickness (~1-2 mm), and then screen printed with silver paste on both sides. After electrode firing, Pt wire leads were connected for the OCV tests. The dense 8YSZ plate was then pressed between an Inconel600 pressing cap (2"x2" and a wall thickness of 0.2") and an alumina base support. The hybrid micas were placed between the 8YSZ plate and the Inconel600 fixture. A schematic drawing of the OCV test fixture is shown in Fig. 1. The OCV measurements were conducted at 800°C after dwelling at the temperature for about 1.5 hours. A low hydrogen-content fuel (2.55~2.71 % H_2/Ar + ~3% H_2O) was used on the anode side (with variable flow rates) and air was used as the oxidant on the cathode side. All the three samples were first fired slowly to 600°C for binder burnout, followed by heating to 830°C for 1 hour and then cooled to the test temperature (800°C) for 2 hours. After the first soaking at 800°C for 2 hours, the samples were furnace cooled to 100°C and started the long-term thermal cycling. The temperature profiles for the thermal cycling are shown in Fig. for sample #2 and #3. Sample #1 was cycled with a different profile, i.e., rapid heating from 100°C to 800°C in 35 minutes, soaking at 800°C for 2 hours, and cooled to 100°C in 5 hours and 25 minutes. For all the three samples, there are 3 cycles per day.

RESULTS AND DISCUSSION

Choice of Low-Hydrogen Fuels for OCV and Thermal Cycle Test

We have chosen a low-hydrogen fuels (2.55-2.71% H_2/Ar) for the long-term thermal cycling and OCV tests in this study. The choice is based on several issues. The first issue is the safety since mica seals are not hermetic seals but with some finite leak rate. During long-term cycling tests, the leak rate may increase to a certain level and presents fire hazard. Use of a low hydrogen fuel with hydrogen content less than 4% would eliminate the safety issue since such a gas is considered nonflammable. The other issue is the reducing environment (or oxygen partial pressure) since pure hydrogen or high hydrogen content fuels are used for actual SOFC stacks. The equilibrium oxygen partial pressure was estimated to be 6.45×10^{-19} atm for a 2.71% H_2/Ar fuel, 1.68×10^{-21} atm for a 50% hydrogen fuel (+3% H_2O), and 4.2×10^{-22} atm for pure hydrogen fuel (97% H_2 + 3% H_2O). It is evident that the use of low hydrogen fuels would still provide a very reducing environment. The theoretical (Nernst) voltage for this gas pair (versus air) when used on an 8YSZ electrolyte would be 0.934 V, a voltage large enough for common DC voltmeter to measure precisely to the sub-mV range. The last issue is the sensitivity to use the low hydrogen fuels in evaluating the thermal cycle stability of the seals with the OCV and thermal cycle tests. The current setup (Fig. 1) does not include the complex internal manifolds as in actual SOFC stacks, and is considered economic, simple, and can provide quick feedback in SOFC seal development. The mica seal was applied only on the anode side and encased (pressed) with the Inconel600 fixture. The cathode side of 8YSZ plate was directly pressed against an alumina support block without any seals in between. The alumina support block has several holes underneath for air flow. The gas pressure at the anode side was maintained slightly higher than the ambient (0.1~0.2 psig) by using a gas bubbler filled with 4~5 inch of water. The leaked fuels would have minimum effect on the cathode air since they would simultaneously react with the surrounding air in the big furnace. On the other hand the air (21 % O_2) leak would have a pronounced effect on the equilibrium oxygen partial at the anode side if a low hydrogen (e.g., 2.71% H_2/Ar) fuel was used as compared to a pure hydrogen fuel (e.g., 100% H_2) simply because there much less hydrogen to react.

Long-Term Thermal Cycling of Hybrid Mica Pressed at 100 psi

In all the OCV tests, the fuel flow rate was chosen at 63 sccm. This flow rate corresponds to an 80% fuel utilization if the cell is to be operated at 0.7 V with a power density of 0.5 w/cm^2 at 800°C and using pure hydrogen (i.e., 97% H_2 + ~3% H_2O). The OCV versus thermal cycles of a 2"x2" 8YSZ plate with the hybrid Phlogopite mica (sample #1) pressed at 100 psi is shown in Fig. 3. This sample has survived more than 711 thermal cycles when thermally cycled between100°C and 800°C. The initial (the first time reaching 800°C) OCV was 0.934 V, exactly matching with the theoretical (Nernst) voltage of the 2.71% H_2/Ar + ~3% H_2O versus air at 800°C. As the thermal cycle continued, the OCV decreased to about 0.92 V in the initial ~50 thermal cycles (Fig. 3B). The OCV tends to stabilize for the rest thermal cycles at ~0.917 V +- 2 mV, corresponding to only 1.8% lower than the Nernst voltage. The drop of OCV in the initial thermal cycles (Fig. 3B) is likely due to the fracture associated with these cycles. For hybrid micas the glass inter-layers will bond strongly to the mating materials and the top several layers of the mica. During thermal cycles, the large CTE mismatch between the Inconel600 fixture (~17 ppm/°C), the mica seal (~11 ppm/°C), and the 8YSZ electrolyte plate (~10.5-11 ppm/°C) would form new fracture (leak path) between the mechanically interlocked flakes through the mica seals. The subsequent frictional wear during following thermal cycles seems to cause no further accumulated damages, and the OCV remained fairly constant. Overall, the current results clearly demonstrated the desired long-term thermal cycle stability of the hybrid Phlogopite mica.

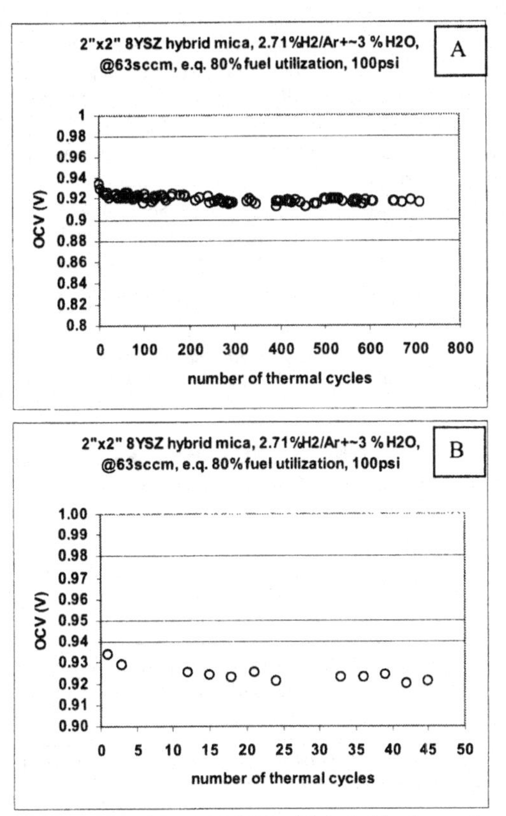

Fig. 3. OCV versus thermal cycles of 2"x2" 8YSZ with hybrid mica seals pressed at 100psi: (A) OCV over the whole range of cycles, (B) OCV of the initial 50 thermal cycles. The fuel was 2.71 % H_2/Ar +~3% H_2O at a flow rate of 63 sccm, equivalent to a 80% fuel utilization of a cell of power density of 0.5 W/cm^2 @800C.

Long-Term Thermal Cycling of Hybrid Mica Pressed at 50 psi

The same hybrid Phlogopite mica (sample #2) was also tested the long-term thermal cycle stability at a lower stress of 50 psi. The results of the measured OCV versus number of thermal cycles are shown in Fig. 4. The initial OCV was 0.928 V, about 6 mV lower than the Nernst voltage (0.934 V). The hybrid mica also showed similar thermal cycle stability as sample #1 that the OCV drop was more evident in the initial ~50 cycles (Fig. 4B), and gradually dropped to 0.906 V after 450 cycles. This sample, however, fractured at cycle #484 and the OCV dropped to 0.72 V. The cause of the 8YSZ electrolyte plate was not clear, and may be attributed to the existing processing flaws, corrosion and penetration of the glass layer, or excess glasses that has been squeezed out from the sealing section during firing under compressive stresses. It is worthy to point that the anode side was maintained at a slightly higher pressure (~0.1 to 0.2 psi) than the surrounding air. One may expect that the leak path would be one-way, i.e., from anode to the surrounding air. Our current setup for OCV test (Fig. 1) does not have internal manifolds like in actual stacks of multiple cells. The fuel leaks into the surrounding air may not affect the oxygen partial pressure at the cathode side, especially the cathode was shielded by an alumina fixture with flowing air. Nevertheless, the low measured OCV at cycle #484 (0.72 V) clearly suggest a major leak of air into the anode side when the 8YSZ electrolyte plate fractured. It needs to be pointed out that the leak rates of common crack in rigid glass seals are often 100-1000 times higher than mica seals. And the crack is highly localized such that a hot-spot will be formed and can lead to total stack failure. Unlike the rigid glass seal, the leak through mica seals is considered evenly distributed since there are continuous network of the voids between discrete mica flakes. As a result, a hot-spot may not form even at the similar total leak rates.

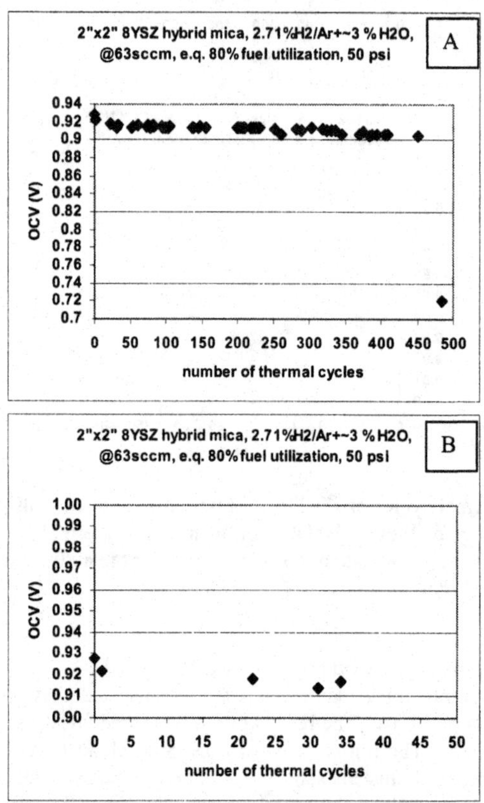

Fig. 4. OCV versus thermal cycles of 2"x2" 8YSZ with hybrid mica seals pressed at 50psi: (A) OCV over the whole range of cycles, (B) OCV of the initial 50 thermal cycles. The fuel was 2.71 % H_2/Ar +~3% H_2O at a flow rate of 63 sccm, equivalent to a 80% fuel utilization of a cell of power density of 0.5 W/cm^2 @800C.

Long-Term Thermal Cycling of Hybrid Glass-Mica Composite Pressed at 26 psi

For actual SOFC stacks the stack size could be 6 inch to 8 inch long, the total area to be sealed could reach, say, 10 in^2. This would require a compressive load of 1000 lbs if the hybrid micas to be pressed at 100psi to satisfy the allowable leak rate (note there are no allowable leak rates set at this time, and often the allowable leak rates will be size and design dependent). Such a high load would definitely present an engineer burden and this drives the need to develop compressive seals of low stresses. In the corresponding paper [11], the authors presented a novel way to further lower the leak rates of the hybrid micas, while maintaining the thermal cycle stability. A concept of "infiltrated" mica was successfully demonstrated with H_3BO_3, Bi-nitrate, and a glass-mica composite that the leak rates can be reduced 10 times as compared to the as-received micas in hybrid form. As pointed out in the paper that H_3BO_3 and Bi-nitrate are not

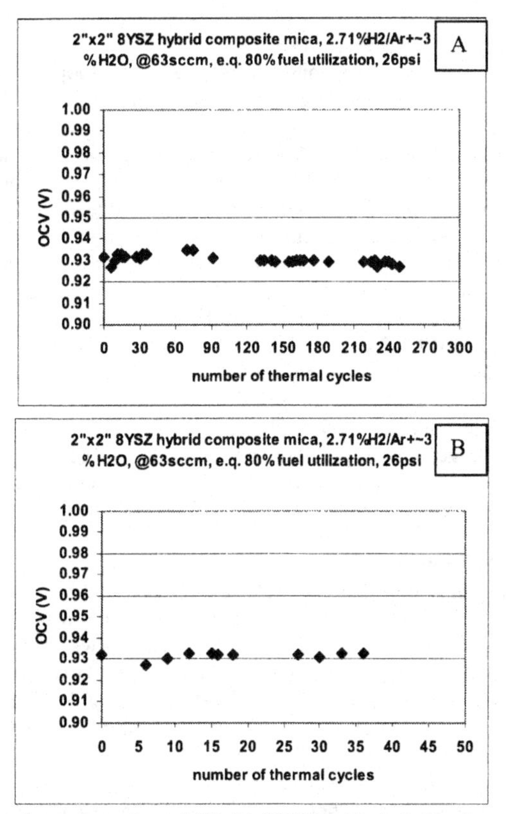

Fig. 5. OCV versus thermal cycles of 2"x2" 8YSZ with hybrid glass-mica composite seal pressed at 26psi: (A) OCV over the whole range of cycles, (B) OCV of the initial 50 thermal cycles. The fuel was 2.71 % H_2/Ar +~3% H_2O at a flow rate of 63 sccm, equivalent to a 80% fuel utilization of a cell of power density of 0.5 W/cm^2 @800C.

considered chemically and thermally stable for long time operation at SOFC environments and temperatures. A glass-mica composite will be durable in these conditions if the glass was carefully selected to meet the SOFC requirements. And a stable Ba-Al silicate glass was chosen to make a glass (20v%)-mica (80v%) composite and tested in the hybrid form for the long-term stability tests (sample #3). The OCV versus thermal cycles was plotted in Fig. 5A, and 5B. The sample showed almost a constant OCV of 0.93 +- 2mV without any distinct OCV drop over the 250 thermal cycles. This sample; however, lost electrical contact after 250 cycles and the test was stopped. Post-test examination has found the Pt-lead wire was broken at a welded section, no fracture of the 8YSZ electrolyte plate or severe damage of the hybrid composite mica. Clearly, the hybrid glass-mica composite seal offered the best thermal cycle stability with almost constant OCV over the entire thermal cycles. The low compressive stress (26 psi) would be more beneficial for SOFC stack designs.

Comparison of Leak rate with the Measured OCVs

The leak rates of the hybrid Phlogopite micas pressed at 100psi at the first cycle were reported to be 0.01-0.04 sccm/cm (standard cubic center meter per outer leak length) at 800°C for a differential gas pressure of 2 psig [8,10]. The leak rate at a differential pressure of 0.1-0.2 psi was not measured for the first cycle, but was expected to be much lower than 0.01-0.04 sccm/cm. The 800°C leak rate of sample #1 at thermal cycle #671 was measured with the ultra-high purity helium initially set at a 2 psig within the sealed anode side through external tubing. The details of the leak test and the rate determination were given in Ref. [6]. By monitoring the pressure drop with time, one can estimate the leak rates through the mica seals. Figure 6A shows the pressure versus time of the leak test of sample #1 at cycle #671. Figure 6B shows the slope (leak rate) of Fig. 6A versus the internal pressure of the ultra-high purity helium. The leak rates at a differential pressure of 0.1-0.2 psi would be 0.012-0.018 sccm/cm at 800°C. For a 2"x2" sample, the total outer leak length would be 2"x4x2.54 cm = 20.32 cm, and the total leak rates would be 0.25-0.37 sccm for a differential pressure of 0.1-0.2 psi. Using this leak rate for air leaks into the anode side with fuel (2.71% H_2/Ar + ~ 3% H_2O) flow rate at 63 sccm, the calculated equilibrium oxygen partial pressure at anode side will be 7.252×10^{-19} atm-8.17×10^{-19} atm. The calculated Nernst voltage would be 0.929 V-0.926 V. Our measured OCV at thermal cycle #671 was 0.920 V, in good agreement with the calculated voltages.

SUMMARY AND CONCLUSIONS

We have conducted the long-term thermal cycle test of three hybrid Phlogopite micas on the 2"x2" dense 8YSZ electrolyte plates. OCV was measured during thermal cycles to assess the stability of these compressive mica seals. The three hybrid micas were pressed at 100 psi, 50 psi, and 26 psi. All the three mica samples showed the desired thermal cycle stability with minimum drop of OCV over thermal cycles as many as 711 cycles. Among the three samples, the hybrid glass-mica composite showed the best thermal cycle stability with OCV remained almost constant during the entire test cycles (250 cycles). The measured OCV also showed good agreement with the calculated OCVs using known leak rates. Overall, the hybrid Phlogopite mica seal has demonstrated the desired thermal cycle stability for solid oxide fuel cells.

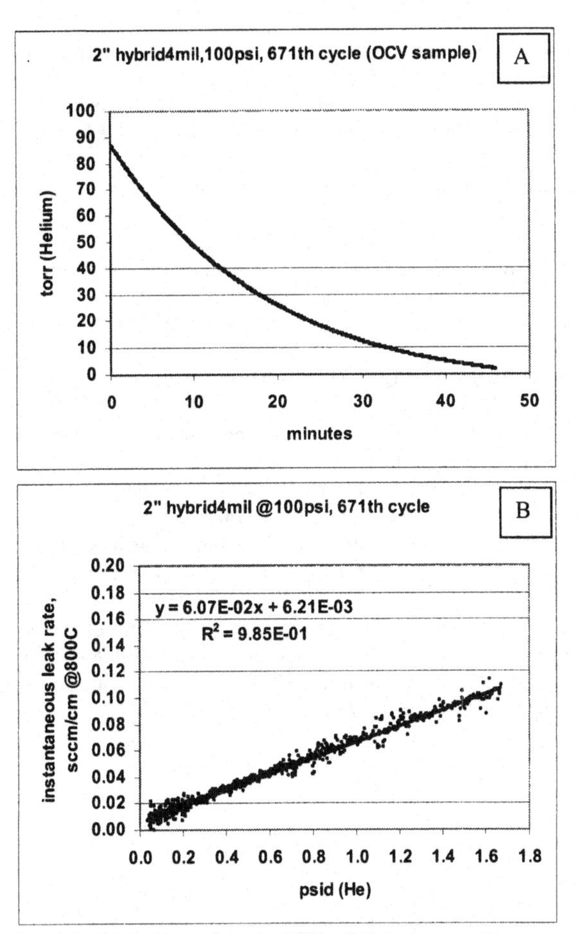

Fig. 6. Leak test of sample #1 at thermal cycle #671. (A) is the actual pressure versus time data, and (B) shows the slope (leak rate) of (A) versus the helium pressure in psid (i.e., pressure difference between the helium and the surrounding air).

ACKNOWLEDGEMENT

The authors would like to thank S. Carlson for SEM sample preparation, and J. Coleman for SEM analysis. This paper was funded as part of the Solid-State Energy Conversion Alliance (SECA) Core Technology Program by the US Department of Energy's National Energy Technology Laboratory (NETL). Pacific Northwest National Laboratory is operated by Battelle Memorial Institute for the US Department of Energy under Contract no. DE-AC06-76RLO 1830.

REFERENCES

[1]T. Yamamoto, H. Itoh, M. Mori, N. Mori, and T. Watanabe, "Compatibility of mica glass-ceramics as gas-sealing materials for SOFC," *Denki Kagaku* **64** [6] 575-581 (1996).

[2]K. Ley, M. Krumpelt, J. Meiser, I. Bloom, *J. Mater. Res.*, **11** 1489 (1996).

[3]N. Lahl, D. Bahadur, K. Singh, L. Singheiser, and K. Hilpert, *J. Electrochem. Soc.*, **149** [5] A607-A614 (2002).

[4]K. S. Weil, J. S. Hardy, and J. Y. Kim, "Use of a Novel Ceramic-to-Metal Braze for Joining in High Temperature Electrochemical Devices," in *Joining of Advanced and Specialty Materials V*, published by the American Society of Metals, 47-55, vol. 5, 2002.

[5]K. S. Weil, J. S. Hardy, and J. Y. Kim, "Development of a Silver-Copper Oxide Braze for Joining Metallic and Ceramic Components in Electrochemical Devices," *Proceedings of the International Brazing and Soldering 2003 Conference*, published by the American Welding Society, 2003.

[6]Y-S Chou, J. W. Stevenson, and L. A. Chick, "Ultra-low leak rate of hybrid compressive mica seals for solid oxide fuel cells," *J. Power Sources*, **112** [1] 130-136 (2002).

[7]S. P. Simner and J. W. Stevenson, "Compressive mica seals for SOFC applications," *J. Power Sources*, **102** [1-2] 310-316 (2001).

[8]Y-S Chou, and J. W. Stevenson, "Mid-term stability of novel mica-based compressive seals for solid oxide fuel cells," *J. Power Sources* **115** [2] 274-278 (2003).

[9]Y-S Chou, and J. W. Stevenson, "Thermal cycling and degradation mechanisms of compressive mica-based seals for solid oxide fuel cells," *J. Power Sources* **112** [2] 376-383 (2002).

[10]Y-S Chou, and J. W. Stevenson, "Phlogopite mica-based compressive seals for solid oxide fuel cells: effect of mica thickness," *J. Power Sources* **124** [2] 473-478 (2002).

[11]Y-S Chou, and J. W. Stevenson, "Infiltrated Phlogopite mica with superior thermal cycle stability as compressed seals for solid oxide fuel cells,"*published in this proceedings..*

ALTERNATIVE METHODS OF SEALING PLANAR SOLID OXIDE FUEL CELLS

K. S. Weil, C. A. Coyle, J. S. Hardy, J. Y. Kim, and G-G. Xia
Pacific Northwest National Laboratory
P. O. Box 999
Richland, WA 99352

ABSTRACT

One of the key limiting issues in designing and fabricating a high performance planar solid oxide fuel cell (pSOFC) stack is the development of the appropriate materials and techniques for hermetically sealing the metal and ceramic components. There are essentially two standard methods of sealing: (1) by forming a rigid joint or (2) by constructing a compressive "sliding" seal. While short-term success has been achieved with both techniques, it is apparent that to meet the long-term operational needs of stack designers, alternative sealing concepts will need to be conceived. Described below are two alternative pSOFC sealing methods that have been developed at Pacific Northwest National Laboratory.

INTRODUCTION

Among solid oxide fuel cell designs, the planar stack (pSOFC) has received growing attention because its compact nature affords high volumetric power density – a design feature of particular importance in transportation applications. With the advent of anode-supported cells that employ thin YSZ electrolytes, these devices can be operated at reduced temperature (700 - 800°C) and still achieve the same current densities exhibited by their high-temperature, thick electrolyte-supported counterparts [1]. The lower operating temperature not only makes it possible to consider inexpensive, commercially available high temperature alloys for use in the stack and balance of plant, but also expands the range of materials that can be considered for device sealing.

Because SOFCs function under an oxygen ion gradient that develops across the electrolyte, hermiticity across this membrane is paramount. In a planar design, this means that the YSZ layer must be dense, must not contain interconnected porosity, and must be connected to the rest of the device with a high temperature, gas-tight seal of the type shown in Figure 1. One of the fundamental challenges in fabricating pSOFCs is how to effectively seal the thin electrochemically active YSZ membrane against the metallic body of the device creating a hermetic, rugged and stable stack. Typical conditions under which these devices are expected to operate and to which the accompanying YSZ-to-metal seal will be exposed include: (1) an average operating temperature of 750°C; (2) continuous exposure to an oxidizing atmosphere on the cathode side and a wet reducing gas on the anode side; and (3) an anticipated device lifetime of 10,000+ hours.

Listed in Table 1 is a generic set of requirements for SOFC seals broken down by functional category. The selection of sealing materials/techniques is closely tied to the specific application in which the stack will be used. Thus it is dependent on a number of design factors, including individual cell and stack materials and geometries, stack assembly sequence, thermal gradients expected across the seal and other stack components, maximum weight and/or volume of the power plant, anticipated external forces, and required heating and/or cooling rate of the device.

One of two techniques is typically used to seal a planar stack: glass joining or compressive

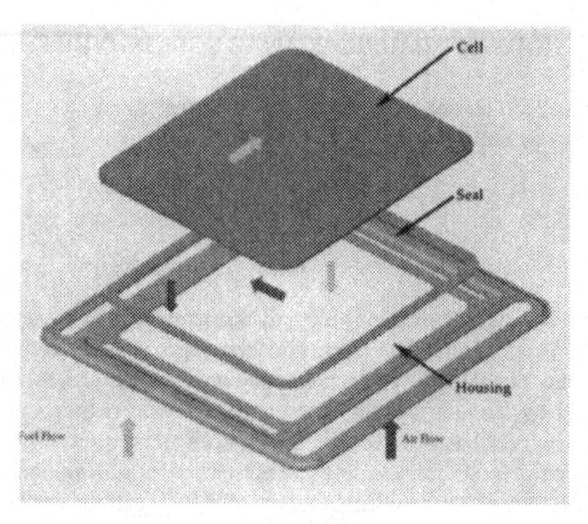

Figure 1 A schematic of the picture frame SOFC cell assembly, with air and fuel flow patterns labeled. The stack is built by joining these cell assemblies one on top of another.

TABLE 1 Functional requirements for pSOFC sealing

Mechanical	Chemical
· Hermetic/marginal leak rate · CTE matching or mitigation of CTE mismatch stresses · Acceptable bond strength or compressive loading requirement (i.e. load frame design) · Resistant to degradation due to thermal cycling/thermal shock · Robustness under external static and dynamic forces*	· Long-term chemical stability under simultaneous oxidizing/wet fuel environments · Long-term chemical compatibility with the adjacent sealing surfaces · Resistance to hydrogen embrittlement
Design/Fabrication	**Electrical**
· Low cost · Facile application/processing · High reliability with respect to achieving initial hermeticity (seal conforms to non-flat substrate surfaces) · Acceptable sealing environment/temperature (i.e. has little effect on the subsequent performance of the stack) · Design flexibility – e.g. allows use of Ni-based alloys in the interconnect*	· Non-conductive (non-shorting configuration)**

* These factors are design specific.
** Glass and ceramic seals are inherently insulating, whereas metal seals are not and the stack design must accommodated this.

sealing. Inherent advantages and limitations are found with each method. For example, glass joining is a cost-effective and relatively simple method of bonding ceramic to metal. However, the final seal is typically brittle and non-yielding, making it particularly susceptible to fracture when exposed to tensile stresses such as those encountered during non-equilibrium thermal events or due to thermal expansion mismatches between the glass and joining substrates [2, 3]. In addition, as the initial glass seal begins to devitrify during the first few hours of high-temperature exposure, its engineered thermal expansion properties change significantly [3], ultimately limiting the number of thermal cycles and the rate of cycling that the stack is capable of surviving. Over time additional problems arise as the sealing material, typically barium aluminosilicate-based, reacts with the chromium- or aluminum oxide scale on the faying surface of the interconnect and forms a mechanically weak barium chromate or celsian phase along this interface [4].

In compressive sealing a compliant, high-temperature material is captured between the two sealing surfaces and compressed, using a load frame external to the stack. Because the sealing material conforms to the adjacent surfaces and is under constant compression during use, it forms a dynamic seal. That is, the sealing surfaces can slide past one another without disrupting the hermeticity of the seal and coefficient of thermal expansion (CTE) matching is not required between the ceramic cell and metallic separator. Unfortunately, this technology remains incomplete due to the lack of a reliable high-temperature sealing material that would form the basis of the compressive seal. A number of materials have been considered, including mica, nickel, and copper, but each has been found deficient for any number of reasons, ranging from oxidation resistance in the case of the metals to poor hermeticity and through-seal leakage with respect to the mica [5]. An additional difficulty is in designing the load frame, as it must be capable of delivering moderate-to-high loads in a high-temperature, oxidizing environment over the entire period of stack operation. Material oxidation and load relaxation due to creep, as well as added expense and additional thermal mass are all issues of concern with this seal design.

We have recently developed an alternative method of ceramic-to-metal brazing specifically for fabricating high temperature solid-state devices such as oxygen generators [6]. Referred to as air brazing, the technique differs from traditional active metal brazing in two important ways: (1) it utilizes a liquid-phase oxide-noble metal melt as the basis for joining and therefore exhibits high-temperature oxidation resistance and (2) the process is conducted directly in air without the use of fluxes and/or inert cover gases. In fact, the strength of the bond formed during air brazing relies on the formation of a thin, adherent oxide scale on the metal substrate. The technique employs a molten oxide that is at least partially soluble in a noble metal solvent to pre-wet the oxide faying surfaces, forming a new surface that the remaining molten filler material easily wets. Potentially, there are a number of metal oxide-noble metal systems that can be considered, including Ag-CuO [7, 8], Ag-V_2O_5, and Pt-Nb_2O_5 [9]. Here we discuss our recent findings on the Ag-CuO system.

A second sealing concept we are developing employs a thin stamped metal foil that is bonded to both sealing surfaces. Unlike a mica gasket, this seal is non-sliding. When properly designed, the foil should readily yield or deform under modest thermo-mechanical loading, thereby mitigating the transfer of these stresses to the adjacent ceramic and metal components. One of the advantages of this sealing concept is that a wide array of alloys could be considered for use in the pSOFC interconnect. Presently the candidate list is limited to those that display good CTE matching with the ceramic cell, namely the chromia scale-forming ferritic stainless steels. If high CTE nickel-based alloys could be used, the mechanical, oxidation, and through-scale electrical properties of the interconnect would be significantly improved [10].

EXPERIMENTAL

Air Brazing

Braze pastes were formulated by mixing the appropriate ratio of copper and silver powders (10μm and 5μm average size, respectively, Alfa Aesar, Inc.) with a standard screen printing binder. The copper oxidizes to CuO in-situ during the brazing operation. Anode-supported bilayers, consisting of NiO-5YSZ as the anode and 5YSZ as the electrolyte, and thin gage FeCrAlY (Fe, 22% Cr, 5% Al, 0.2% Y) were employed as the model SOFC electrolyte membrane/structural metal system in our study. Disc-shaped bilayer coupons were fabricated by traditional tape casting and co-sintering techniques and measured 25mm in diameter by 600 μm in thickness, with an average electrolyte thickness of ~8 μm. As-received 300μm thick FeCrAlY sheet was cut into washer-shaped specimens measuring 4.4 cm in diameter with a concentric 1.5 cm diameter hole and cleaned with acetone prior to joining. The components for the rupture test specimen are shown in Figure 2(a).

Joining was conducted by applying a concentric 24mm ring of braze paste to the FeCrAlY washer using an automated pressure-driven dispenser. After allowing the paste to dry, the bilayer was placed YSZ-side down onto the washer and dead-loaded with 25g of weight. The assembly was heated in air at 20°C/min to 1050°C (brazing can be conducted between 970 and 1100°C). and held at temperature for 15min before furnace cooling to room temperature. The hermeticity and ultimate strength of the seal were measured using the rupture test rig shown in Figures 2(b) and (c). Hermeticity is determined by pressurizing the sample to a value less than that required for failure and measuring whether the pressure subsequently decays as a function of time. Seal strength is determined by slowly pressurizing the sample to the point of failure. Thermal cycle testing was conducted by heating the specimens in air at a rate of 75°C/min to 750°C, holding at temperature for ten minutes, and cooling to ≤70°C in forty minutes before re-heating under the same conditions. A minimum of six specimens was tested for each test condition. Microstructural analysis was conducted on polished cross-sectioned samples using a JEOL JSM-5900LV scanning electron microscope (SEM) equipped with an Oxford energy dispersive X-ray analysis (EDX) system.

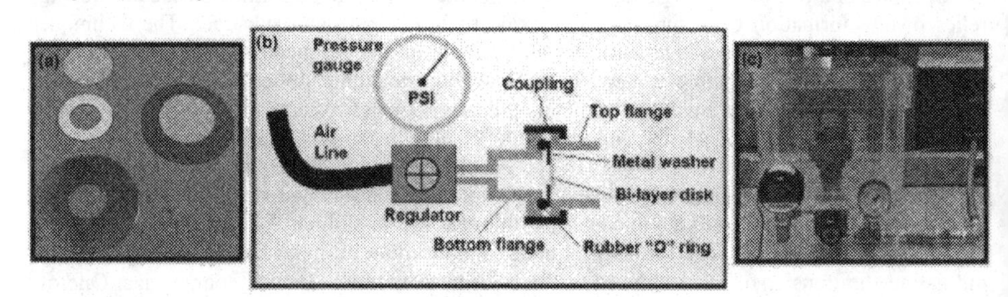

Figure 2 (a) Components of rupture strength test specimen: from the bottom, the FeCrAlY washer, brazing filler metal foil ring, and bi-layer disk. To the right, the assembled test specimen. (b) A schematic and (c) a photo of the rupture test rig.

Bonded Compliant Seal

A number of high temperature alloys can be considered for use as the foil membrane in the bonded compliant seal concept. As part of a proof-of-concept study, our initial materials screening analysis focused on four key properties: high oxidation resistance, low stiffness, high ductility, and low cost. Based on these factors, we chose a commercial alumina-forming ferritic steel as the foil membrane: DuraFoil (22% Cr, 7% Al, 0.1%La+Ce, bal. Fe, manufactured by Engineered Materials Solutions, Inc.). Supplied as 50μm thick sheet, the DuraFoil was sheared into 3 cm x 3 cm samples, annealed in vacuum at 900°C for 2hrs, and stamped into cap-shaped washers using a die designed specifically for this purpose. The stamped foils were ultrasonically cleaned in soap and water, and then flushed with acetone to remove the lubricant from the stamping operation.

Each foil washer was bonded to a 6.2mm thick Haynes 214 washer, with an outside diameter of 4.4cm and an inside diameter of 1.5cm, using BNi-2 braze tape (Wall Colmonoy, Inc.). An alumina-scale forming nickel-based superalloy, Haynes 214 displays excellent oxidation resistance at temperatures in excess of 1000°C, but also exhibits an average CTE of 15.7 μm/m·K, which is almost 50% higher than that of the anode-supported bilayer (CTE = 10.6 μm/m ·K). Fabrication of the specimen was completed by air brazing the top side of the stamped foil to the YSZ side of a 25mm diameter bilayer disc using a Ag-4mol% CuO paste at the conditions described previously. The specimens were characterized via rupture and thermal cycle testing and subsequently analyzed by SEM and EDX.

RESULTS

Air Brazing

Shown in Figure 3(a) is a cross-sectional micrograph of a typical brazed joint, which measures ~90μm thick and extends from the YSZ electrolyte to the alumina scale on the FeCrAlY. Although the 4mol% CuO composition used wets the YSZ quite well, no reaction zone was found at the YSZ/braze interface. The interface however is decorated by an array of widely dispersed CuO precipitates. Small equiaxed crystals of CuO are observed in the bulk region of the joint, surrounded by an essentially pure silver matrix. Between the braze filler metal and the FeCrAlY, a well-defined reaction zone ~2μm thick is observed. EDX analysis indicates that this region is a mixture of two phases, $CuO{\cdot}Al_2O_3$ and $2CuO{\cdot}Al_2O_3$. Adjacent to the reaction zone is an ~0.5μm thick alumina scale that forms on the FeCrAlY during brazing. If a non-optimized braze composition is employed in joining, porosity can develop along the joint interfaces due to regions of non-wetting, as seen in the low-CuO braze joint shown in Figure 3(b). Alternatively, a high-CuO containing braze joint exhibits the microstructure observed in Figure 3(c), which is characterized by a thick CuO layer that envelopes each interface. As shown in Figure 4, sessile drop wetting experiments and corresponding rupture strength testing indicate that optimal Ag-CuO compositions for brazing YSZ to FeCrAlY lie in a narrow band centered approximately at 4mol% CuO.

Results from rupture testing are shown in Figure 5. All of the tested specimens were found to be hermetic and combined thermal cycle/rupture testing indicated no failure in any of the seals out to 40 cycles, the maximum number thermal cycles considered in our study. Failure occurred instead within the ceramic disc, as seen in the inset, suggesting this as the weakest component in the current specimen configuration. In comparison, similar specimens fabricated with a conventional barium calcium aluminosilicate sealing glass (of the type used in Reference [10]) displayed an average initial strength of 20.6 psi, which degraded to nearly zero after ten thermal cycles. Failure in these glass sealed specimens appears to initiate at the interface between the sealing glass and the scale on the stainless steel. Given these results, we have begun sealing full-size planar stack components. Initial studies indicate that the air brazing process is readily scalable and that leak-tight seals can be formed in parts as large as 15cm x 10cm. We intend to continue thermal cycle testing of these components, as well as exposure testing and full-scale stack studies.

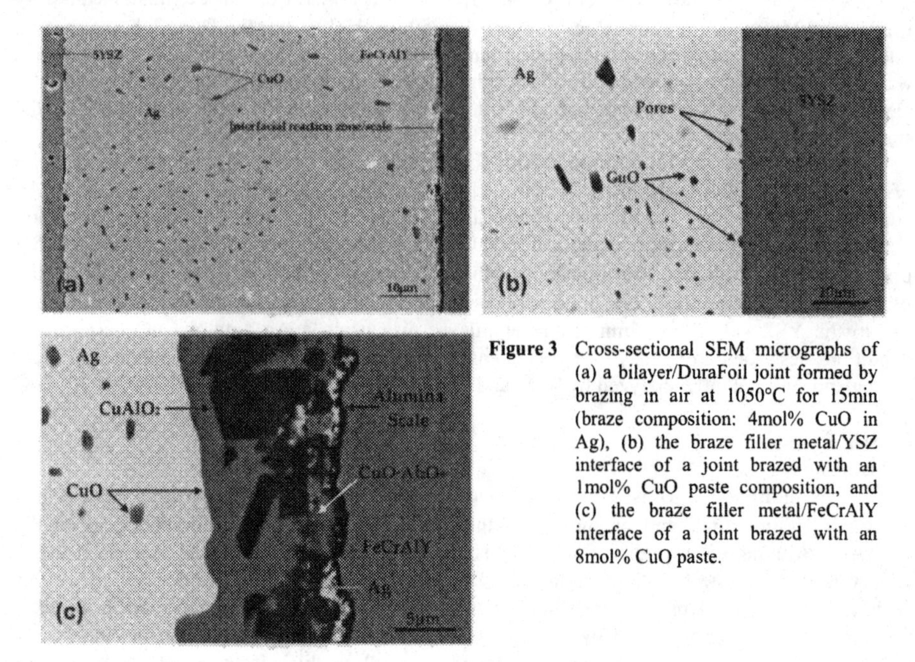

Figure 3 Cross-sectional SEM micrographs of (a) a bilayer/DuraFoil joint formed by brazing in air at 1050°C for 15min (braze composition: 4mol% CuO in Ag), (b) the braze filler metal/YSZ interface of a joint brazed with an 1mol% CuO paste composition, and (c) the braze filler metal/FeCrAlY interface of a joint brazed with an 8mol% CuO paste.

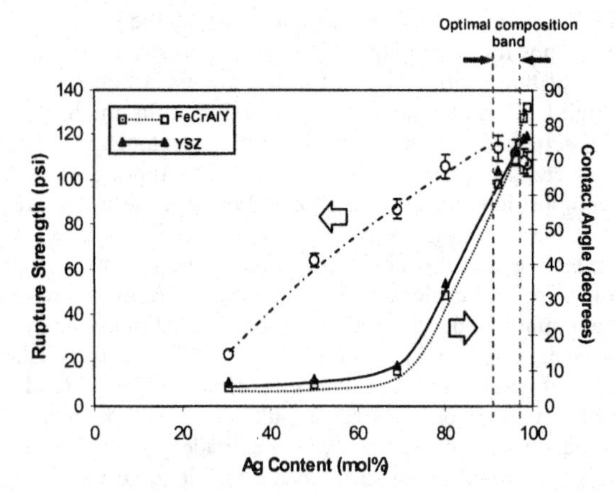

Figure 4 Room temperature rupture strength and contact angle (on YSZ and FeCrAlY) as a function of Ag content.

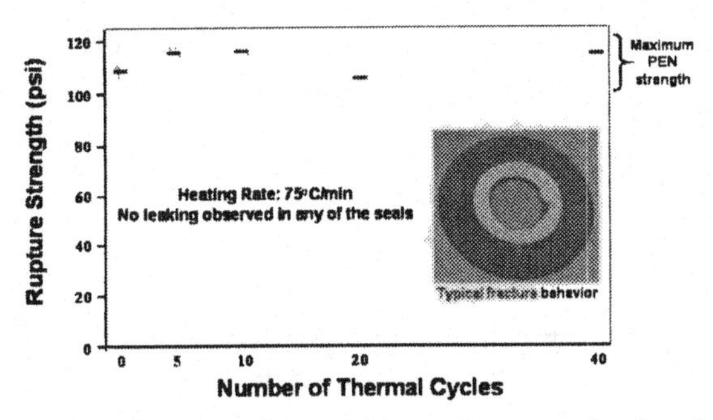

Figure 5 Rupture strength of the air brazed joint specimens as a function of rapid thermal cycling. Inset: typical mode of failure observed in the rupture specimens, within the center of the ceramic disc.

Bonded Compliant Seal

Shown in Figure 6(a) is a composite cross-sectional micrograph of a bonded compliant seal specimen. The sample was well sealed, as determined by hermeticity testing conducted prior to metallographic analysis. The entire seal is approximately 1.1mm thick, although it is expected that this can be readily reduced simply by altering the geometry of the DuraFoil stamping. In this sample, the BNi-2 braze has caused the outer periphery of the foil to curl due to a mismatch in CTE between the two materials. While this does not degrade the strength of the rupture specimen, it could potentially affect the performance of the seal in an actual stack; an issue that will need to be addressed. On the ceramic side of the seal, the CuO-Ag braze appears to form a robust joint between the YSZ and the alumina scale of the DuraFoil. Note that the braze is thicker toward the center of the specimen. No reaction zone is observed at the YSZ/braze interface, however a 10 – 15μm thick zone forms on the DuraFoil due to reaction between the Al_2O_3 scale and the CuO in the braze. The dominant product is the mixed oxide phase $2CuO \cdot Al_2O_3$. A conceptual drawing of the seal in an actual application is shown in figure 6(b). Sealing occurs around the gas manifold holes, which are incorporated into both the ceramic cell and the separator plate. Results from rupture testing are shown in Figure 7. Each specimen was found to be hermetic up to the maximum pressure (60 psi) tested during initial leak testing. More

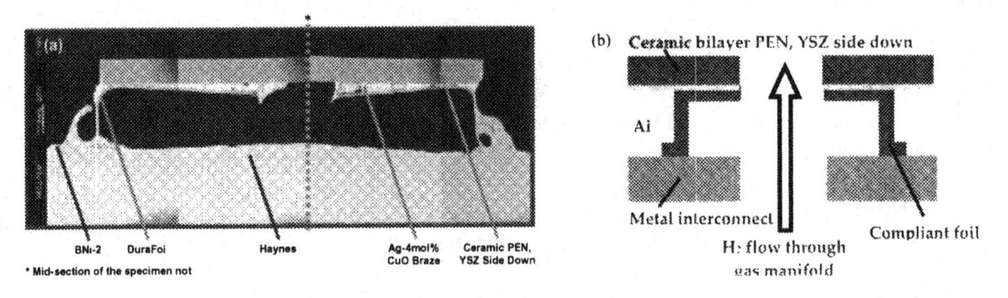

Figure 6 (a) A composite cross-sectional micrograph of an as-joined bonded compliant seal rupture test specimen. (b) A schematic of the bonded compliant sealing concept in use in pSOFC gas manifold sealing.

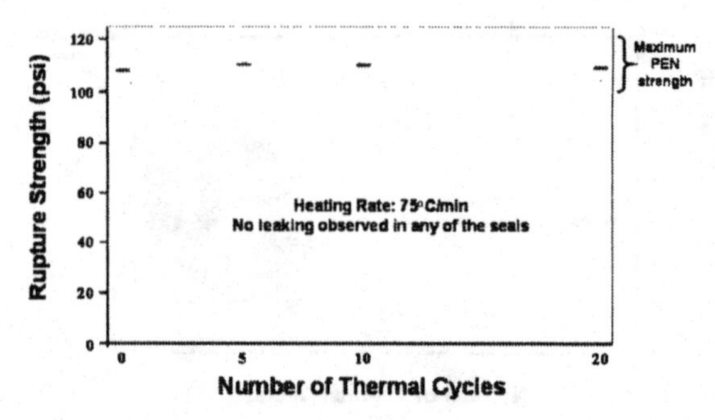

Figure 7 Rupture strength of the bonded compliant seal specimens in the as-joined and as-cycled conditions.

extensive pressure testing up to the point of PEN rupture indicated no failure in any of the seals, even in the specimens that underwent as many as twenty rapid thermal cycles. As with the air brazed specimens above, failure occurred in the ceramic disc.

CONCLUSIONS

Planar SOFCs continue to hold much promise for efficient, high density power generation. To fulfill this promise, robust sealing technologies must be developed that can meet the functional requirements of both stack designers and manufacturers. Glass joining has proven to be an effective means of sealing stacks for short- and moderate-term (<3000hr, few thermal cycles) testing, but questions have been raised concerning the long-term durability and thermal cycling performance of these seals. Compressive sealing remains in its infancy, burdened by problems with through-seal leakage. At Pacific Northwest National Laboratory, we are developing two alternative methods of sealing planar stacks: air brazing and bonded compliant sealing. Results from recent studies demonstrate hermetic YSZ-to-stainless steel joints can be formed in both cases, and these joints remain leak-free and well bonded after aggressive thermal cycle testing. Continuing work will investigate the long-term durability of the seals, as well as the potential for scale-up and prototypic use in demonstration stacks.

ACKNOWLEDGMENTS

The authors would like to thank Nat Saenz, Shelly Carlson, and Jim Coleman for their assistance. This work was supported by the U. S. Department of Energy as part of the SECA and the FE-ARM programs. The Pacific Northwest National Laboratory is operated by Battelle Memorial Institute for the United States Department of Energy (U.S. DOE) under Contract DE-AC06-76RLO 1830.

REFERENCES

1. B.C.H. Steele, A. Heinzel (2001) Materials for fuel-cell technologies, *Nature*, **414**(X) 345-52.
2. K. Eichler, G. Solow, P. Otschik, W. Schaffrath (1999) BAS ($BaO{\cdot}Al_2O_3{\cdot}SiO_2$) glasses for high temperature applications, *J. Eur. Cer. Soc.*, **19**(6-7) 1101-4.

3. Z.G. Yang, K.S. Weil, D.M. Paxton, K.D. Meinhardt, J. W. Stevenson (2003) Considerations of glass sealing solid oxide fuel cell stacks, *in:* J. E. Indacochea, J. N. DuPont, T. J. Lienert, W. Tillmann, N. Sobczak, W. F. Gale, M. Singh (Eds.) Joining of Advanced and Specialty Materials V, ASM International, Materials Park, OH, 40-48.
4. Z.G. Yang, K.S. Weil, K.D. Meinhardt, J.W. Stevenson, D.M. Paxton, G.-G. Xia, D.-S. Kim (2002) Chemical compatibility of barium-calcium-aluminosilicate base sealing glasses with heat resistant alloys, *in:* J. E. Indacochea, J. N. DuPont, T. J. Lienert, W. Tillmann, N. Sobczak, W. F. Gale, M. Singh (Eds.) Joining of Advanced and Specialty Materials V, ASM International, Materials Park, OH, 116-24.
5. S.P. Simner, J.W. Stevenson (2001) Compressive mica seals for SOFC applications, *J. Power Sources,* **102** (1-2) 310-6.
6. J.S. Hardy, J.Y. Kim, K.S. Weil (in press) Joining mixed conducting oxides using an air-fired electrically conductive braze, *J. Electrochem. Soc.*
7. Z.B. Shao, K.R. Liu, L.Q. Liu, H.K. Liu, S. Dou (1993) Equilibrium phase diagrams in the systems PbO-Ag and CuO-Ag, *J. Am. Cer. Soc.*, **76** (10) 2663-4.
8. A.M. Meier, P.R. Chidambaram, G.R. Edwards (1995) A comparison of the wettability of copper-copper oxide and silver-copper oxide on polycrystalline alumina, *J. Mater. Sci.*, **30** (19) 4781-6.
9. R.S. Roth, J.R. Dennis, H.F. McMurdie, eds. (1987) *Phase Diagrams for Ceramists, Volume VI*, The American Ceramic Society, Westerville, OH.
10. Z.G. Yang, K.S. Weil, D.M. Paxton, J.W. Stevenson (2003) Selection and evaluation of heat-resistant alloys for SOFC interconnect applications, *J. Electrochem. Soc.*, **150** (9) A1188-201.

INFILTRATED PHLOGOPITE MICAS WITH SUPERIOR THERMAL CYCLE STABILITY AS COMPRESSIVE SEALS FOR SOLID OXIDE FUEL CELLS

Yeong-Shyung Chou and Jeffry W. Stevenson
K2-44, Materials Department
Pacific Northwest National Laboratory
P O. Box 999,
Richland, WA 99352

ABSTRACT

Thermal cycle stability is one of the most stringent requirements for sealants in solid oxide fuel cell stacks. The sealants have to survive several hundreds to thousands of thermal cycles during lifetime operation in stationary and transportation applications. Recently, researchers at the Pacific Northwest National Laboratory have developed a novel method to improve the thermal cycle stability of mica seals. The main concept of the method is to infiltrate the mica flakes with wetting or liquid forming materials in order to reduce leak path connectivity from 3-D to 2-D and to achieve good thermal cycle stability with low leak rates. Leak rates were determined for mica seals infiltrated with H_3BO_3 or bismuth nitrate; the concept was also tested on a glass-mica composite. The infiltrated mica seals were also tested with cells, for which open circuit voltages were measured and compared to the Nernst voltages.

INTRODUCTION

One of the most challenging tasks facing planar solid oxide fuel cell (SOFC) developers is the need for reliable sealing technology. Seals that will offer long-term stability in the high temperature SOFC environment (oxidizing and reducing) and maintain their integrity during hundreds to thousands of thermal cycling are required. Currently, there are three primary approaches for SOFC seal development: glass (or glass-ceramic) seals [1-4], metallic brazes [5-6], and compressive seals [7-11]. The use of compressive seals offers a unique advantage over the other approaches in that a stringent matching of the coefficient of thermal expansion (CTE) of the various SOFC stack components is not necessary. Currently, the development of compressive seals is primarily focused on mica-based seals. Chou and Stevenson recently developed novel hybrid mica seals through which the 800°C leak rates were reduced 2 to 3 orders of magnitude by adding a glass or metal interlayer between the mica materials and adjacent stack components [7-8]. These Muscovite mica-based seals also showed good thermal stability when held for about 500 hours in air or reducing environments at elevated temperatures [10]. The cleaved Muscovite mica sheets, however, suffered a rapid increase in leak rate during thermal cycling, due to through-thickness cracks and wear damage from the repeated cycles; in some cases, the leak rates tended to stabilize after 20~30 thermal cycles [11]. The objective of the present study is to develop a compressive mica seal offering constant, low leak rates during repeated thermal cycling. In this paper, we present a novel sealing approach based on commercially available Phlogopite mica papers infiltrated with forming materials. Results of leak rate and open circuit voltage (OCV) testing of these infiltrated mica seals are reported and compared with results for glass-ceramic seals.

EXPERIMENTAL

Materials

The mica used in this study is a commercially available Phlogopite mica paper (McMaster-Carr, Atlanta, GA). The mica paper is composed of discrete mica flakes (Fig. 1A). Due to the anisotropic morphology of the mica flakes, they are highly oriented, with their basal planes (cleavage plane) overlapping with each other (Fig. 1B). Two Phlogopite mica papers were used in this study. One is about 0.004" thick and contains 3~5% of organic binders (mica-A). The other is about 0.003" thick and contains no binders (mica-B). Both materials have similar color and surface textures. Prior to infiltration, mica-A was first heat-treated at 700°C for 4 hours to remove the binders. Mica-B was used as-received. For the infiltration, two chemicals were used: H_3BO_3 (99.5% Alfa Aesar, MA) and $Bi(NO_3)_3 \cdot 5H_2O$ (98%, Alfa Aesar, MA). De-ionized water was used to make saturated H_3BO_3 or bismuth nitrate solutions. The boric acid solution was heated to 70~90°C for infiltration. Mica-A discs (about 1.5 inch in diameter) were immersed into the saturated H_3BO_3 solution for ~2 minutes, removed, and dried at ~50°C. Bismuth nitrate solutions were also made in a similar manner and the infiltration was conducted on mica-B with a pipette instead of immersing the mica discs into the solution. Bismuth nitrate did not form a clear aqueous solution but instead decomposed to form white precipitates which left a thick coating on the micas. In addition to these two approaches, a glass-mica composite was also fabricated by mixing mica flakes (90v%) with Ba-Al silicate glass powder (10v%) in water to which a small amount (2 wt%) of binder was added. The slurry was poured into a plastic mold lined with Mylar film and dried at ambient conditions before cutting into the desired shape and size for leak testing.

Leak Rate Tests with Thermal Cycling

Leak rate tests of the infiltrated mica discs were conducted by pressing them between an Inconel600 pipe (OD=1.3" and ID=1.0") and an alumina substrate (2"x2"). A compressive load of 100 psi was maintained throughout the whole test, including the heating and cooling cycles. The leak rates were determined using ultra-high purity helium at a gauge pressure of 2 psi. Thermal cycling was conducted between 100°C and 800°C in air. Details of the leak test, leak rate determination, temperature profiles for thermal cycling, and mid-term stability tests are given elsewhere [7, 10, 11].

Open Circuit Voltage Tests

In order to more fully assess the sealing capability of the mica seals, open circuit voltage (OCV) tests were conducted on in-house made 8-YSZ electrolyte plates with various mica seals. 2"x2" dense 8-YSZ plates were prepared by slip casting 8-YSZ powders (TOSOH, Zirconia TZ-8Y, Japan), followed by sintering at 1450°C for 2 hours. The sintered plates were machined to the desired size and thickness (~1-2 mm), and then screen printed with silver paste electrodes on both sides. After electrode firing, Pt wire leads were connected for the OCV tests. The dense 8-YSZ plate was then pressed between an Inconel pressing cap and an alumina base support. The infiltrated micas were placed between the 8-YSZ and the Inconel fixture.

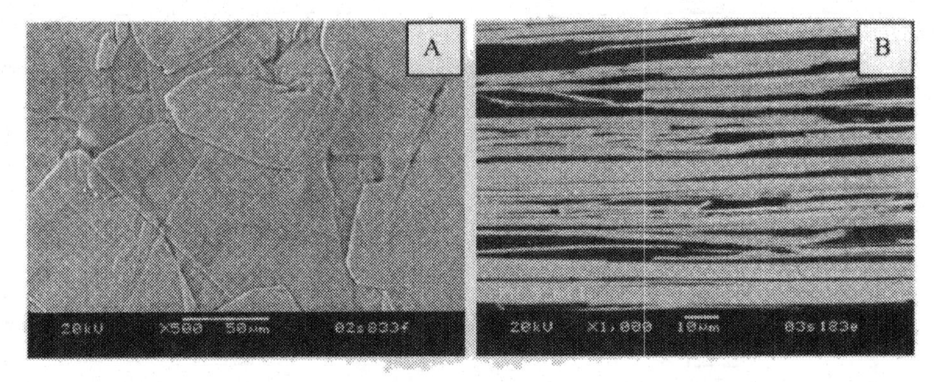

Fig. 1. Morphology of the commercial Phlogopite mica paper : (A) top surface and (B) cross-section.

A schematic drawing of the OCV test fixture is shown in Fig. 2. The OCV measurements were conducted at 800°C after dwelling at temperature for about 2 hours. A low hydrogen-content fuel (2.55~2.71 % H_2/ balance Ar + ~3% H_2O) was used on the anode side (with variable flow rates) and air was used as the oxidant on the cathode side.

RESULTS AND DISCUSSION

Concept of Infiltrated Mica and Leak Paths

In previous work on naturally cleaved monolithic Muscovite mica we learned that the leak rates would increase rapidly during initial thermal cycles, and tended to saturate after 20-30 thermal cycles [10]. Post-test characterization revealed through-thickness cracks and wear damage in the thermally cycled Muscovite mica, possibly due to the very low CTE of the Muscovite (~7 ppm/°C) compared to the two mating materials in the tests: Inconel600 pipe (CTE ~17 ppm/°C) and alumina substrate (~8-9 ppm/°C). In subsequent work, Phlogopite mica, which has a higher CTE (~11 ppm/°C), was found to offer improved leak rate stability over repeated thermal cycles [11]. However, the observed leak rates may still be considered high for SOFC stacks (though the allowable leak rates for SOFC stacks have not yet been determined), and therefore it is desirable to further reduce the leak rates. The concept of infiltrated mica was proposed after examining cross-sections of the mica papers (Fig. 1B), which contain voids between the discrete mica flakes. These connected voids create significant leak paths due to their 3-dimensional connectivity. The leak paths could be minimized if these voids were partially filled with a wetting material (such as a glass or a material at its melting point) to break up the 3-D connectivity. It was anticipated that new fracture surfaces formed during thermal cycling would create leak paths which were limited to the newly-formed fracture plane (with 2-D connectivity). This concept is illustrated schematically in Fig. 2 where Fig. 2A shows multiple leak paths (dotted lines) in the as-received mica paper as

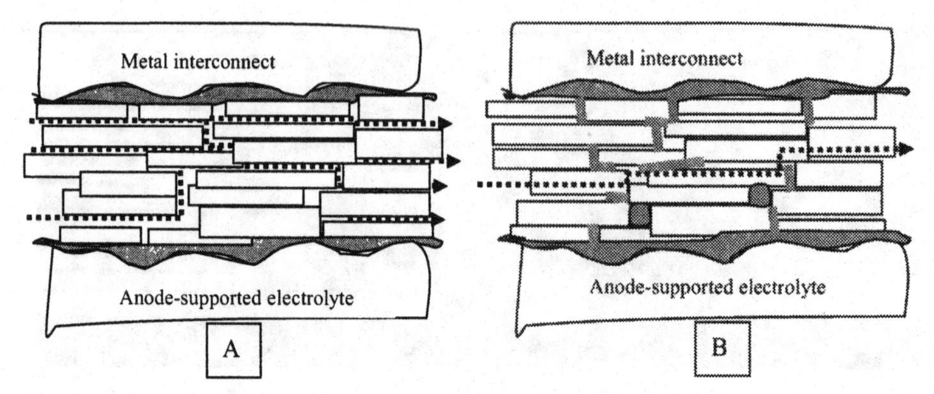

Fig. 2. Schematic showing the concept of infiltrated mica in which the continuous voids between discrete mica flakes (rectangles) are filled or blocked with an infiltrated wetting or compliant phase (gray) such that upon thermal cycling a desirable 2-D leak path (dotted line in B) will form instead of a 3-D path (dotted lines in A).

compared to the single leak path (dotted line) in the infiltrated mica (Fig. 2B) (in both cases, the mica is shown in "hybrid" form, i.e., sandwiched between two glass interlayers).

Choice of the Infiltration Materials

To test the "infiltrated" mica concept, two chemicals were selected: boric acid (H_3BO_3) and bismuth nitrate ($Bi(NO_3)_3 \cdot 5H_2O$). Boric acid converts to B_2O_3 upon heating and was expected to form a wetting liquid at elevated temperatures (M.P. ~ 450°C). Similarly, Bismuth nitrate converts to Bi_2O_3 in air. Bismuth oxide has a higher M.P. (~813°C), but was expected to be deformable at 800°C, leading to the possibility that it could be squeezed into the voids between flakes when compressive stresses were applied. It needs to be mentioned that these two chemicals may not offer satisfactory long-term stability in SOFC operating environments due to high volatility. However, they were considered to be appropriate choices for these proof-of-concept experiments. A third approach was to use a glass-mica composite with 10v% glass and 90v% mica flakes premixed. In this approach the glass powder were already distributed between mica flakes, and was expected to form a wetting liquid at the voids at elevated temperature and stress.

800°C Leak Rate of H_3BO_3-infiltrated Phlogopite Micas versus Thermal Cycling

The results of the 800°C thermal cycle leak rate tests of the H_3BO_3-infiltrated Phlogopite micas (in hybrid form, i.e., a glass (Ba-Al-Silicate) layer was inserted above and below the infiltrated mica) are shown in Fig. 3. Previous results for the same mica without infiltration (also in hybrid form under the same compressive stress) are also included [10,11]. It is interesting to note that the leak rates of the infiltrated mica decreased abruptly with an increasing number of thermal cycles. The leak rates were ~ 2.3×10^{-2} sccm/cm initially, ~ 5.5×10^{-4} sccm/cm after 15 thermal cycles, and ~1.3×10^{-4}

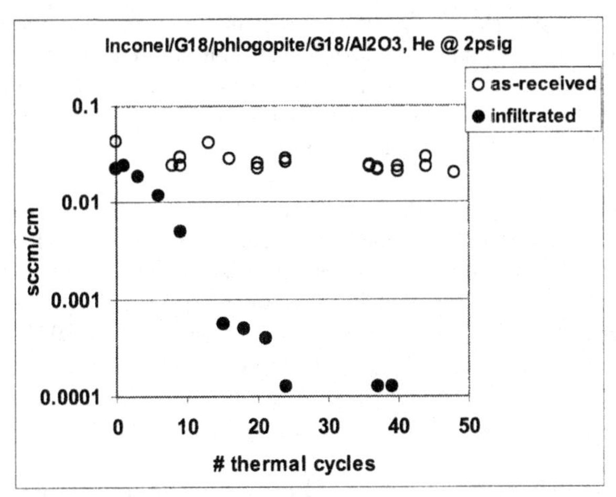

Fig. 3. Effect of thermal cycling on the 800°C leak rates of the Phlogopite mica in the as-received and the H_3BO_3-infiltrated forms. The mica was pressed between an Inconel pipe and an alumina substrate at 100 psi in air.

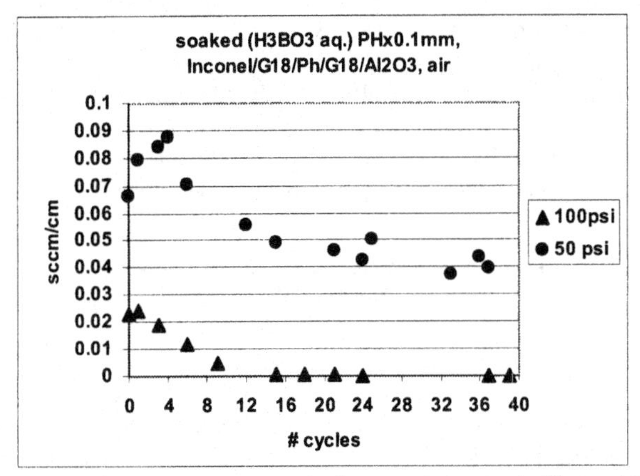

Fig. 4. Effect of compressive stress on the leak rate of H_3BO_3-infiltrated Phlogopite mica during thermal cycling.

sccm/cm after 23 cycles. The reduction of leak rate is about 2 orders of magnitude. For the as-received mica, the leak rate was ~ $4.3x10^{-2}$ sccm/cm initially, and remained fairly constant (within the range of 2.1 ~ $3.0x10^{-2}$ sccm/cm) during the following 48 thermal cycles. It is evident that the presence of the B_2O_3 must have substantially blocked the leak paths between the mica flakes. It is not clear why the leak rates of the infiltrated micas were higher in the beginning than they were after ~23 thermal cycles. One possible cause is that the initial distribution of B_2O_3 was not uniform within the mica paper, but instead was more highly concentrated on the outer surfaces of the mica samples. During the test period, B_2O_3 may have gradually penetrated into the mica paper, steadily blocking the continuous leak paths between discrete mica flakes. Post-test observation showed that the mica sample could be easily detached from the Inconel test fixtures, indicating that the B_2O_3 did not react with the mica (an alumina silicate mineral) to form a significant amount of glass at the interface.

Effect of Compressive Stresses on the Leak Rate of H_3BO_3-infiltrated Phlogopite Micas

The effect of thermal cycling on the leak rate of H_3BO_3-infiltrated Phlogopite micas was also tested at a lower compressive stress (50 psi). The leak rate data are plotted in Fig. 4 along with data of higher compressive stress (100 psi). It appeared that lower compressive stresses did not yield low leak rates; however, the 800°C leak rates still showed the desired thermal cycle stability in that the leak rates remained fairly constant over ~36 thermal cycles. One possible cause for the higher leak rates is that the voids between mica flakes were larger such that there was not enough glass to fill them.

Effect of CTE mismatch on the Leak Rate of Bismuth nitrate-Infiltrated Phlogopite Micas

The leak tests of the bismuth nitrate infiltrated micas were conducted in a different way than the previous H_3BO_3-infiltrated mica tests. To examine the effect of CTE mismatch, the bismuth nitrate infiltrated micas were tested in three different metal-metal couples, instead of the Inconel600-alumina substrate couple. The infiltrated micas were pressed between an Inconel600 pipe and one of the three different metal support plates: Inconel600, Haynes230, and SS430. The average CTE (RT to 800°C) of Inconel600, Haynes 230, and SS430 are ~17 ppm/°C, ~15 ppm/°C, and ~12.5 ppm/°C, respectively. Inconel600 and Haynes230 are superalloys with superior thermal, mechanical, and oxidation resistance properties relative to the ferritic stainless steel (SS430). However, due to their very high CTEs, they can only be considered as SOFC interconnect candidates if compliant seals (such as compressive seals) are used. SS430 has a good

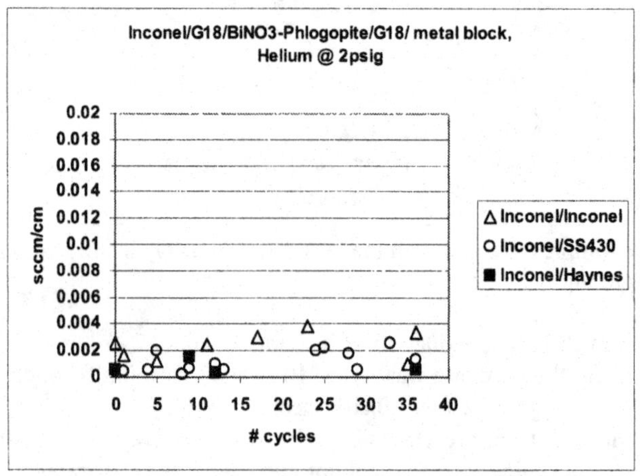

Fig. 5. Effect of thermal cycling on the 800°C leak rates of bismuth nitrate infiltrated Phlogopite mica pressed between 3 different metal couples. The mica was pressed between an Inconel pipe and a metal substrate of Inconel, Haynes230, and SS430.

CTE match with typical anode-supported SOFCs, but suffers severe oxidation at 800°C. Our tests of the infiltrated mica in these three metal couples therefore covered a wide range of CTE mismatch over which other sealing approaches (e.g., glass seals and brazes) are not likely to be suitable.

The leak rates versus thermal cycling of the infiltrated mica in the three metal couples are shown in Fig. 5. It is interesting to note that the leak rates are much lower for all three metal couples, i.e., less than 4×10^{-3} sccm/cm in Fig. 5 (the as-received mica has leak rates of $2.1 \sim 4.3 \times 10^{-2}$ sccm/cm). Leak rates through the bismuth nitrate infiltrated mica appeared to be independent of the wide range of CTE mismatches among the three metal couples. In addition, the leak rates were rather insensitive to the number of thermal cycles (the fluctuation of leak rates during the 36 cycles is likely due to ambient temperature fluctuations).

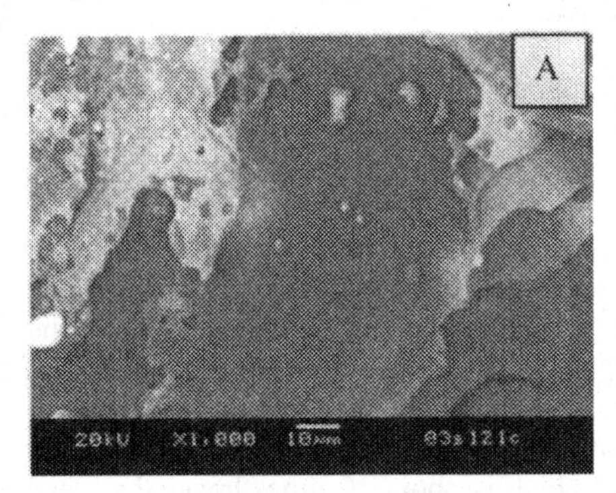

Fig. 6. Scanning electron micrograph showing the fracture surface of the bismuth nitrate infiltrated mica after 37 thermal cycles in air at 100 psi (A). Figure B shows the cross-sectional view of infiltrated mica which was not thermally cycled. In both pictures the white phase is Bi_2O_3 and the gray phase is mica.

Compared to the H_3BO_3-infiltrated micas, the bismuth nitrate infiltrated micas showed low leak rates throughout the cycling testing instead of a gradual decrease with thermal cycles. This behavior is likely due to the presence of excess bismuth nitrate on the mica surfaces as well as the thinner mica used (Mica-B, ~75 μm). It was found that bismuth nitrate ($Bi(NO_3)_3 \cdot 5H_2O$) did not form a stable aqueous solution but decomposed to form $BiONO_3 \cdot H_2O$ white precipitates. After infiltration and drying, a coating of the precipitate remained on the mica discs. In post-cycling microstructural characterization, it was found that the mica could be easily detached from the test fixtures, indicating that the micas were sufficiently intact to be easily cleaved along their basal planes. The penetration of Bi_2O_3 into the mica paper was evident on the top surface and the cross-section micrographs (the white phase in Fig. 6A and 6B). The gray phase is the mica in both SEM pictures.

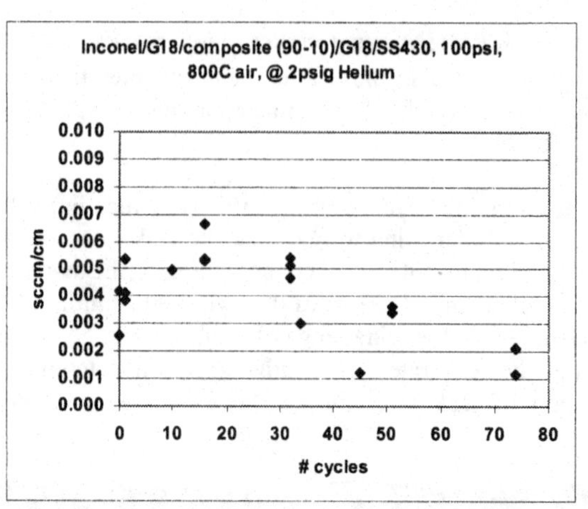

Fig. 7 Effect of thermal cycling on the 800°C leak rates of the glass (10v%)-mica (90v%) composite seal. The mica was pressed between an Inconel600 pipe and a SS430 substrate at 100 psi in air.

800°C Leak Rate of Glass-Mica Composite versus Thermal Cycling

As mentioned earlier H_3BO_3 and bismuth nitrate are not considered chemically and thermally stable in the SOFC environments (oxidizing, humid, and reducing), so the infiltration concept was further tested on a glass(10v%)-mica(90v%) composite. The leak rate versus thermal cycle plot is shown in Fig. 7. It is evident that the concept was proven successful in lowering leak rates and maintaining thermal cycle stability. After 75 cycles, the leak rates were 1-2x10^{-3} sccm/cm, which is about 10 times lower than lead rates observed for the as-received mica (Fig. 3).

Open Circuit Voltage Test of Various Seals

In addition to the leak rate measurements, open circuit voltage (OCV) tests were also performed. OCV tests provide information regarding the effectiveness of the seal by allowing a comparison of the observed cell voltage with the theoretical (Nernst) voltage. If the measured OCV is lower than the theoretical voltage, it often indicates leakage. OCV tests were conducted on bismuth nitrate infiltrated micas, and, for a comparison, on a glass seal and on plain Phlogopite mica (without infiltration and not in the hybrid form). In these tests, only the fuel side was sealed and a low-hydrogen fuel (2.71% H_2/ balance Ar with ~3% H_2O) was used. Although actual fuels for SOFC operation are much higher in hydrogen content, the use of 2.71% H_2/Ar fuel offered advantages of safety and sensitivity to leaks from the air side into the fuel side. The results of the OCV tests at various fuel flow rates are plotted in Fig. 8. The Nernst voltage of this fuel gas versus air at 800°C was calculated to be 0.934 V. It is evident that the OCVs of the glass seal and the Bi-infiltrated micas were in agreement with the theoretical voltages (within +/- 0.5%) indicating the sealing capability of the infiltrated mica was as good as the sealing glass. In addition, the effect of the fuel flow rate appeared to be very minute as the OCV only changed by a few mV when the flow rate was increased from 25 sccm to about 200 sccm, consistent with the fact that the leak rates through these infiltrated micas are very low. On the other hand, the plain mica showed OCVs much lower than the Nernst voltage (the

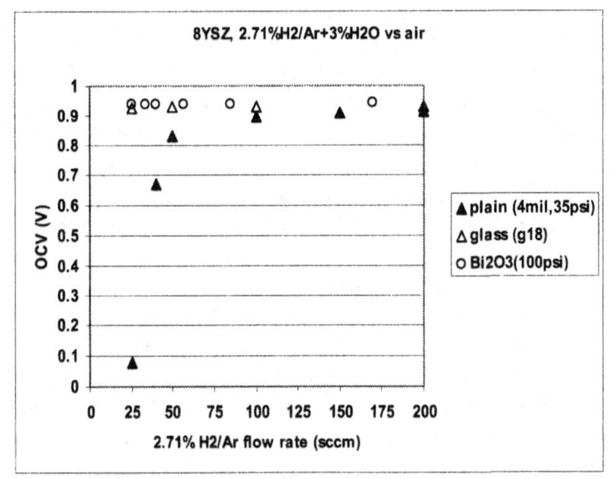

Fig. 8 OCV of 2"x2" 8-YSZ electrolyte at 800°C with various seals: a glass seal, a bismuth nitrate infiltrated mica seal, and a plain mica seal. The fuel is 2.71% H_2/Ar + ~3% H_2O.

plain mica was pressed at a much lower stress of 35 psi, and was expected to have appreciable leak rates (1-10 sccm/cm) [8]. In addition, the OCVs of the plain mica increased with increasing fuel flow rate. This suggests a major leak of air into the fuel side and a dilution effect at higher fuel flow rates.

SUMMARY AND CONCLUSIONS

A novel "infiltrated" mica seal concept was proposed and successfully demonstrated. The "infiltrated" mica paper seal was tested with Phlogopite mica paper and a wetting or melt-forming agent (H_3BO_3 and bismuth nitrate). The results for H_3BO_3-infiltrated mica showed a continued decrease in leak rate during thermal cycling, and very low rates (<$5x10^{-4}$ sccm/cm) were obtained after ~15 thermal cycles. The results for bismuth nitrate-infiltrated mica exhibited constant leak rates of 1~4 $x10^{-3}$ sccm/cm over 36 thermal cycles and showed no dependence on CTE mismatches. The concept was also demonstrated on a glass-mica composite with low leak rates after 75 thermal cycles. OCV values equal or close to the Nernst voltage were obtained for the infiltrated micas. These results clearly demonstrate the viability of the concept of "infiltrated" micas as candidates for compressive seals for solid oxide fuel cells.

ACKNOWLEDGEMENT

The authors would like to thank S. Carlson for SEM sample preparation, and J. Coleman for SEM analysis. This paper was funded as part of the Solid-State Energy Conversion Alliance (SECA) Core Technology Program by the US Department of Energy's National Energy Technology Laboratory (NETL). Pacific Northwest National

Laboratory is operated by Battelle Memorial Institute for the US Department of Energy under Contract no. DE-AC06-76RLO 1830.

REFERENCES

[1]T. Yamamoto, H. Itoh, M. Mori, N. Mori, and T. Watanabe, "Compatibility of mica glass-ceramics as gas-sealing materials for SOFC," *Denki Kagaku* **64** [6] 575-581 (1996).

[2]K. Ley, M. Krumpelt, J. Meiser, I. Bloom, *J. Mater. Res.,* **11** 1489 (1996).

[3]P. Larsen, C. Bagger, M. Morgensen, J. Larsen, "Stacking of planar SOFCs," pp 69-78, in M. Dokiya, O. Yamamoto, H. Tagawa, S. Singhal (Eds.), Solid Oxide Fuel Cells-IV, Vol. 69, Electrochemical Society, Pennington, NJ, PV 95-1, 1995.

[4]N. Lahl, D. Bahadur, K. Singh, L. Singheiser, and K. Hilpert, "Chemical interactions between aluminosilicate base sealants and the components on the anode side of solid oxide fuel cells," *J. Electrochem. Soc.,* **149** [5] A607-A614 (2002).

[5]K. S. Weil, J. S. Hardy, and J. Y. Kim, "Use of a Novel Ceramic-to-Metal Braze for Joining in High Temperature Electrochemical Devices," pp. 47-55 in *Joining of Advanced and Specialty Materials V*, published by the American Society of Metals, 2002.

[6]K. S. Weil, J. S. Hardy, and J. Y. Kim, "Development of a Silver-Copper Oxide Braze for Joining Metallic and Ceramic Components in Electrochemical Devices," *Proceedings of the International Brazing and Soldering 2003 Conference*, published by the American Welding Society, 2003.

[7]Y-S Chou, J. W. Stevenson, and L. A. Chick, "Ultra-low leak rate of hybrid compressive mica seals for solid oxide fuel cells," *J. Power Sources*, **112** [1] 130-136 (2002).

[8]S. P. Simner and J. W. Stevenson, "Compressive mica seals for SOFC applications," *J. Power Sources*, **102** [1-2] 310-316 (2001).

[9]Y-S Chou, and J. W. Stevenson, "Mid-term stability of novel mica-based compressive seals for solid oxide fuel cells," *J. Power Sources* **115** [2] 274-278 (2003).

[10]Y-S Chou, and J. W. Stevenson, "Thermal cycling and degradation mechanisms of compressive mica-based seals for solid oxide fuel cells," *J. Power Sources* **112** [2] 376-383 (2002).

[11]Y-S Chou, and J. W. Stevenson, "Phlogopite mica-based compressive seals for solid oxide fuel cells: effect of mica thickness," *J. Power Sources* **124** [2] 473-478 (2002).

Lithium Ion Batteries

MANGANESE OXIDE CATHODES FOR TRANSPORTATION APPLICATIONS

Arumugam Manthiram and Youngjoon Shin*
The University of Texas at Austin
Materials Science and Engineering
1 University Station C2200
Austin, TX 78712-0292, USA

ABSTRACT

The doubly substituted spinel lithium manganese oxides such as $LiMn_{2-y-z}Ni_yLi_zO_4$ show excellent performance characteristics that are appealing for electric and hybrid electric vehicle applications. They exhibit excellent cyclability at elevated temperatures, high rate capability, good storage performance, low irreversible capacity (IRC) loss, and low electrical resistance. The superior performance compared to the conventional $LiMn_2O_4$ spinel oxide is attributed to a smaller lattice parameter difference Δa between the two cubic phases formed around $(1-x) \approx 0.3 - 0.5$ in $Li_{1-x}Mn_{2-y}Ni_yLi_zO_4$ during the charge-discharge process and the consequent low microstrain or lattice strain. Additionally, the electrochemical properties are found to bear a clear relationship to the initial Mn valence and the initial lattice parameter. The excellent electrochemical performance coupled with the inexpensive and environmentally benign nature of Mn and good structural and chemical stabilities may make optimized spinel oxide compositions viable candidates for lithium ion cells suitable for transportation applications.

INTRODUCTION

Growing environmental concerns have created an enormous interest in electric and hybrid vehicles (EV and HEV).[1] However, some issues with the power supplying systems such as batteries or fuel cells and driving systems need to be addressed to make the EV and HEV technologies successful. For example, the problems with the cost and energy density of the power supplying systems need to be overcome. From an energy density point of view, lithium ion batteries are attractive for EV and HEV applications as they offer higher energy density compared to other rechargeable systems. Unfortunately, the high cost and toxicity of the currently used layered lithium cobalt oxide cathodes remain to be an impediment to develop the lithium ion technology for transportation applications. Lithium manganese oxides are appealing in this regard as manganese is inexpensive and environmentally benign.[2] However, the spinel lithium manganese oxide that has been investigated over the years is plagued by severe capacity fade during cycling particularly at elevated temperatures. To improve the cycling performance, cationic substitutions such as Li, Mg, and Al for Mn have been pursued, and they have led to better performance.[3-5] In addition, several mechanisms including Jahn-Teller distortion and manganese dissolution have been proposed to be the source of capacity fade.[6,7] However, the mechanisms proposed could explain only some parts of the capacity fade. With an aim to develop a better understanding of the capacity fading mechanisms and to utilize the understanding to design and develop high performance spinel manganese oxide compositions

---

*Present Address: LG Chem, Ltd./Research Park, 104-1 Moonji-Dong Yuseong-Gu Daejeon, 305-380, Korea

with acceptable performance, we have focused on a systematic investigation of a number of cation substituted spinel manganese oxides. By investigating several singly and doubly substituted spinel manganese oxides, a clear relationship between capacity fade and some intrinsic materials characteristics such as initial lattice parameter, initial manganese valence, and the lattice parameter difference Δa between the two cubic phases formed during the discharge-charge process is found. The understanding is used to develop doubly substituted compositions such as $LiMn_{2-y-z}Ni_yLi_zO_4$ that exhibit superior capacity retention, high rate capability, excellent storage performance, and low irreversible capacity (IRC) loss in the first cycle.

EXPERIMENTAL

The singly substituted $LiMn_{2-y}M_yO_4$ (M = Li, Al, Ti, Co, Ni, and Cu and $0 \le y \le 0.2$) and doubly substituted $LiMn_{2-y-z}M_yLi_zO_4$ (M = Al, Ti, Fe, Co, Ni, Cu, and Ga, $0 \le y \le 0.1$, and $0 \le z \le 0.1$) samples were synthesized by firing mixtures of Li_2CO_3 and Mn_2O_3 with Al_2O_3, TiO_2, Fe_3O_4, Co_3O_4, NiO, CuO, or Ga_2O_3 at 800 °C for 48 h in air. All samples were characterized by X-ray diffraction. The electrochemical performance of the $LiMn_{2-y}M_yO_4$ and $LiMn_{2-y-z}M_yLi_zO_4$ cathodes were evaluated with CR2032 coin cells using metallic lithium anode and 1 M $LiPF_6$ in ethylene carbonate (EC) and diethyl carbonate (DEC) electrolyte. The cathodes were fabricated by mixing 75 wt% of transition metal oxide with 20 wt% acetylene black and 5 wt% of polytetrafluoroethylene (PTFE) binder, rolling the mixture into thin sheets of about 0.2 mm thick, and cutting into circular electrodes of 0.64 cm^2 area. Electrochemical data were collected between 4.3 and 3.5 V at various rates between C/10 and 4C at room temperature and 60 °C. Electrode resistances were investigated with a current interruption technique during charging. After the current was interrupted, the voltage relaxation was monitored with time, and the electrode resistance was calculated from the initial iR drop obtained by extrapolation.

Samples for monitoring the lattice parameter variations with lithium content were obtained by chemical lithium extraction, which was carried out by stirring the transition metal oxide powder with an acetonitrile solution of the oxidizing agent NO_2BF_4 at room temperature for 2 days under argon atmosphere using a Schlenk line followed by washing the products with acetonitrile and drying under vacuum at room temperature.[8] Lattice parameters were determined by Rietveld analysis using the DBWS-9411 PC program.[9] The lithium contents in the chemically delithiated $Li_{1-x}Mn_{2-y-z}M_yLi_zO_4$ samples were determined by atomic absorption spectroscopy.

RESULTS AND DISCUSSION

Power supplying systems for EV and HEV applications need to satisfy different requirements compared to those for small portable electronic devices such as cell phones and laptop computers. In the case of small portable devices, the energy density (Wh/L) is the most important factor since portable devices have limited space for batteries. However, in the case of transportation applications, the weight of batteries is more important than the volume since the weight is directly related to the range of vehicles. In addition, power and storage performance of batteries should be considered since the role of batteries is to help the engine to climb a hill and accelerate rapidly. Also, a longer life for batteries is preferred compared to the life of the vehicles. Considering these points, we have focused on an assessment of the cycle life, rate capability, electrode resistance, and storage performance of the manganese oxide cathodes.

Figure 1 compares the cyclability data of the singly substituted $LiMn_{2-y}M_yO_4$ and doubly substituted $LiMn_{2-y-z}M_yLi_zO_4$ samples at 60 °C. As seen in Figure 1, all the cation-substituted samples show much better capacity retention compared to the unsubstituted $LiMn_2O_4$ both at room temperature and at elevated temperatures although some of them exhibit lower capacities.[10] Especially, $LiMn_{1.9}Ni_{0.05}Li_{0.05}O_4$ loses only 7.4 % of its initial capacity in 50 cycles at 60 °C compared to more than 50 % for $LiMn_2O_4$. Additionally, it is observed that for the same degree of cationic substitution, some of the doubly substituted samples exhibit better capacity retention than the singly substituted samples. For example, $LiMn_{1.9}Ni_{0.05}Li_{0.05}O_4$ exhibits lower capacity loss (3.2 %) than the singly substituted $LiMn_{1.9}Li_{0.1}O_4$ (4.2 %) and $LiMn_{1.9}Ni_{0.1}O_4$ (9.6 %) in 50 cycles at room temperature.[10]

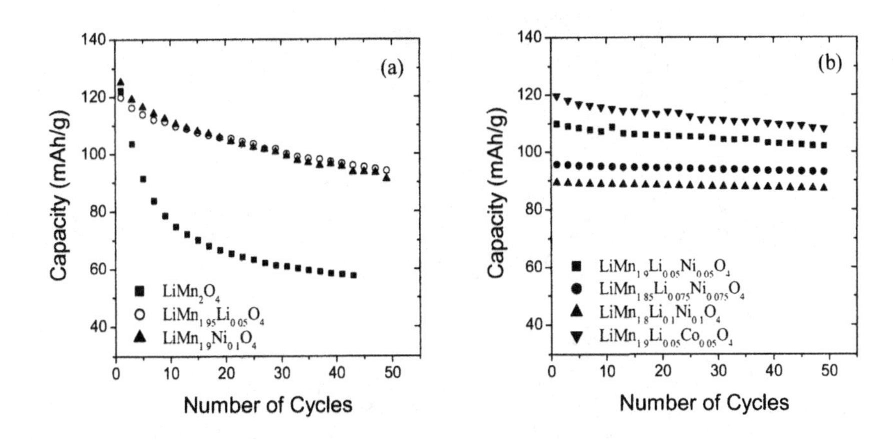

Figure 1: Comparison of the cycling performances of (a) singly and (b) doubly substituted spinel manganese oxides with that of $LiMn_2O_4$ at C/5 rate (~ 0.2 mA/cm^2) between 4.3 and 3.5 V at 60 °C.

Rate capabilities of the cation substituted spinel oxides showing excellent cyclability were investigated between C/10 and 4C rates at room temperature. The experiments were carried out by discharging the CR2032 coin cells down to 3.5 V at different rates of C/10, C/5, C/2, 1C, 2C, and 4C after first charging up to 4.3 V at the same rate of C/5. The discharge profiles of the doubly substituted $LiMn_{2-y-z}Ni_yLi_zO_4$ samples at different rates are compared with those of the unsubstituted $LiMn_2O_4$ and the currently used $LiCoO_2$ in Figure 2; the $LiCoO_2$ sample was synthesized by a solid state reaction at 900 °C for 24 h. The cation-substituted samples generally exhibit better rate capabilities compared to $LiMn_2O_4$. Especially, the doubly substituted samples show better rate capability than the singly substituted samples. For example, $LiMn_{1.85}Ni_{0.075}Li_{0.075}O_4$ retains 98 % of its capacity on increasing the rate from C/10 to 4C compared to 47 and 91 %, respectively, for $LiMn_2O_4$ and $LiCoO_2$. Furthermore, the doubly substituted $LiMn_{1.85}Ni_{0.075}Li_{0.075}O_4$ sample retains 84 and 57% of its capacity on increasing the rate from C/10 to 10C and 20C respectively.

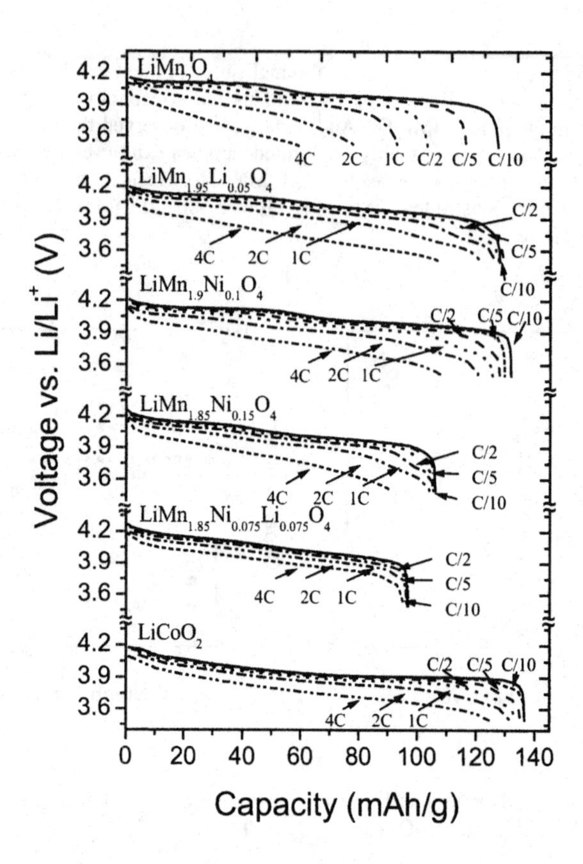

Figure 2: Comparison of the discharge profiles illustrating the rate capabilities of $LiMn_{2-y-z}Ni_yLi_zO_4$ and $LiCoO_2$.

Another important factor in achieving high power is the low electrode resistance. For EV and HEV applications, the low resistance is an important factor since the batteries may be discharged in a short time (pulse discharge) rather than fully discharged. Figure 3 compares the electrode resistance measured by the current interruption technique of selected cation substituted spinel manganese oxides with that of $LiMn_2O_4$ containing different amounts of conductive carbon. While the electrodes fabricated with 20 wt% conductive carbon do not show significant differences in electrode resistance (Figure 3a), those fabricated with 3 wt% carbon do exhibit some differences (Figure 3b). With 3 wt% conductive carbon, the $LiMn_2O_4$ electrode exhibits a much higher electrode resistance of 2,000 to 10,000 ohm compared to the cation substituted samples (< 2,000 ohm) as seen in Figure 3b. Especially, the electrode resistance of $LiMn_{1.85}Ni_{0.075}Li_{0.075}O_4$ with 3 wt% conductive carbon is less than 1/10 of that of $LiMn_2O_4$ at 1-x

≤ 0.5. Furthermore, the electrodes of the cation substituted spinel oxides could be charged even without adding conductive carbon. It appears that the cation substitution improves the electrical conductivity of the bulk samples, which might help to improve the power performance.

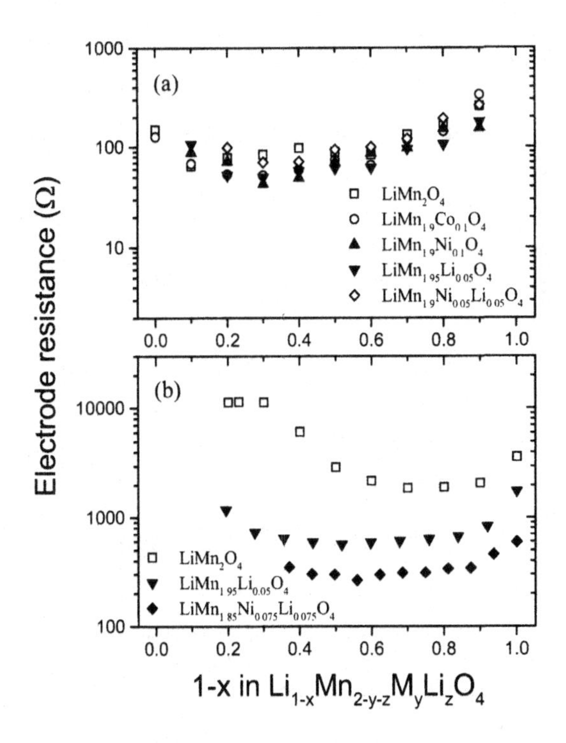

Figure 3: Comparison of the electrode resistances of selected cation-substituted spinel manganese oxides during the first charge: (a) electrodes with 20 % conductive carbon and (b) electrodes with 3 % conductive carbon.

The storage performances were evaluated by subjecting the coin cells to one charge-discharge cycling between 4.3 and 3.5 V at room temperature, discharging to various depths of discharge (DOD) in the second cycle, storing at 60 °C for 7 days at various DOD, discharging the remaining capacity in the second cycle, subjecting to the third charge, and evaluating the discharge capacity in the third cycle at room temperature. Figure 4 gives the % capacity recovered in the third discharge cycle after storing at various DOD compared to the discharge capacity in the first cycle for $LiMn_2O_4$ and $LiMn_{1.85}Ni_{0.075}Li_{0.075}O_4$. While $LiMn_2O_4$ loses a significant amount of capacity (10 – 40 %) after storage, $LiMn_{1.85}Ni_{0.075}Li_{0.075}O_4$ retains > 95 % of its initial capacity, illustrating excellent storage characteristics. Thus the excellent storage properties also make the doubly substituted spinel manganese oxide appealing for transportation applications. In addition, the irreversible capacity (IRC) loss in the first cycle, which is one of the important performance factors since only the reversible capacity can be utilized in secondary batteries, was much smaller for the doubly substituted spinel oxides compared to that for $LiMn_2O_4$.[10]

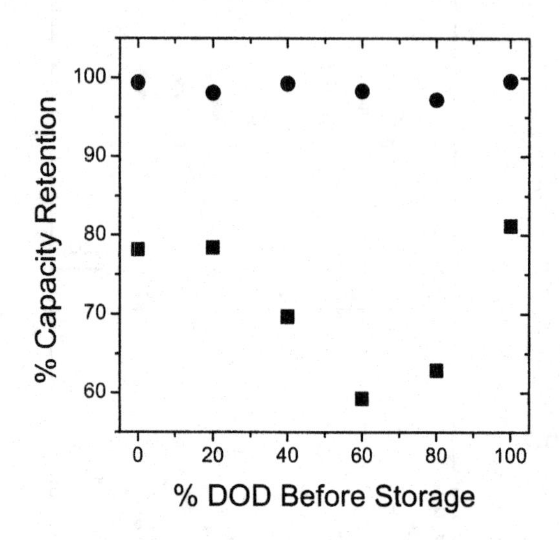

Figure 4: Comparison of the % capacity retention after storing at 60 °C for 7 days at different depth of discharge (DOD). ■: $LiMn_2O_4$ and ●: $LiMn_{1.85}Ni_{0.075}Li_{0.075}O_4$. See the text for experimental details in evaluating the storage properties.

To understand the reason for the improved performance of the doubly substituted spinel oxides, we focused on several intrinsic sample parameters related to the crystal structure since Jahn-Teller distortion and Mn dissolution could not fully account for the remarkable improvement in the electrochemical performance of the doubly substituted spinel oxides.[10] For example, the capacity fade was correlated to the initial lattice parameter values and the initial manganese valence of the samples. The manganese valences were calculated assuming an oxygen content value of 4.0, which is reasonable as the samples were synthesized at 800 °C followed by cooling slowly in air. Figure 5 correlates the % capacity fade to initial lattice parameter and Mn valence. As seen in Figures 5a and 5b, the % capacity fade seems to decrease with decreasing initial lattice parameter and increasing initial Mn valence, but only in the region with < 10 % capacity fade. Figure 6 shows the variations of the initial lattice parameter and the % capacity fade with initial Mn valence. The initial lattice parameter decreases with increasing Mn valence since the amount of smaller Mn^{4+} ions would increase with increasing Mn valence (Figures 5a, 5b, and 6a). Interestingly, the % capacity loss versus Mn valence seems to have two regions with a boundary around a Mn valence of 3.58+ (Figures 5b and 6b). Above an initial Mn valence of 3.58+, the capacity fade decreases sharply and remains small.

However, a number of samples that have a constant initial Mn valence of around 3.55+ and initial lattice parameter values of around 8.24 Å exhibit a wide variation in capacity fade (10-40 %) as indicated by nearly horizontal lines in Figures 5a and 5b for % capacity fade of > 10 %. While one could account for the better cyclability of the samples with higher Mn valence (> 3.58+) to be due to the suppression of Jahn-Teller distortion, the horizontal nature of the data points for capacity fades of > 10 % in Figures 5a and 5b suggest that Jahn-Teller distortion may not be the only factor influencing the capacity fade. Additionally, no systematic correlation of the capacity fade was found with the degree of manganese dissolution.

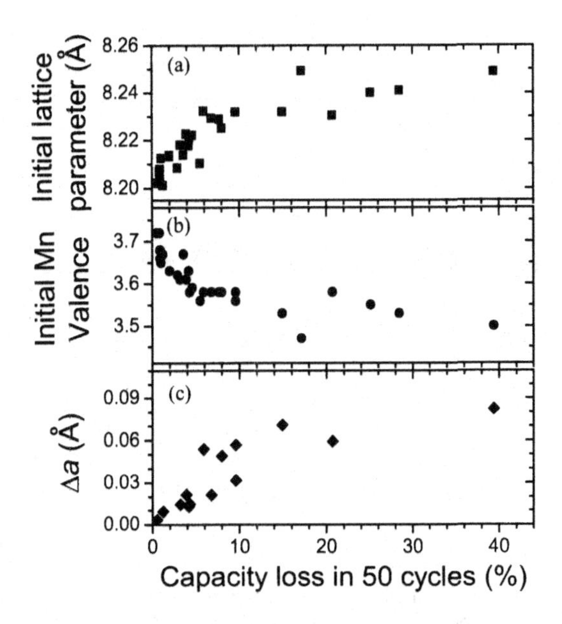

Figure 5: Correlation of the % capacity fade in 50 cycles at room temperature to the (a) initial lattice parameter, (b) initial Mn valence, and (c) lattice parameter difference Δa between the two cubic phases formed in the two-phase region during the charge-discharge process of the spinel oxides.

With an aim to develop a better understanding of the capacity fade mechanism, we then focused on the cubic to cubic phase transition occurring in the 4 V region of the $LiMn_2O_4$ spinel oxides. In this regard, we have examined the evolution of the X-ray diffraction patterns with lithium content by chemically extracting lithium from the spinel oxides and the lattice parameter difference between the two cubic phases that coexist around (1-x) ≈ 0.3 – 0.5 in the spinel $Li_{1-x}Mn_2O_4$ system. Although $LiMn_2O_4$ and the singly substituted spinel oxides show two well-separated, distinct peaks around (1-x) ≈ 0.3 – 0.5 corresponding to the two cubic phases that have a larger difference in lattice parameters, the doubly substituted samples show only broad peaks without a clear splitting due to close lattice parameters for the two phases. Despite the broad peaks for the $Li_{1-x}Mn_{2-y-z}M_yLi_zO_4$ samples in the two-phase region, the reflections could be resolved by Rietveld refinement to obtain the lattice parameters for the two cubic phases.

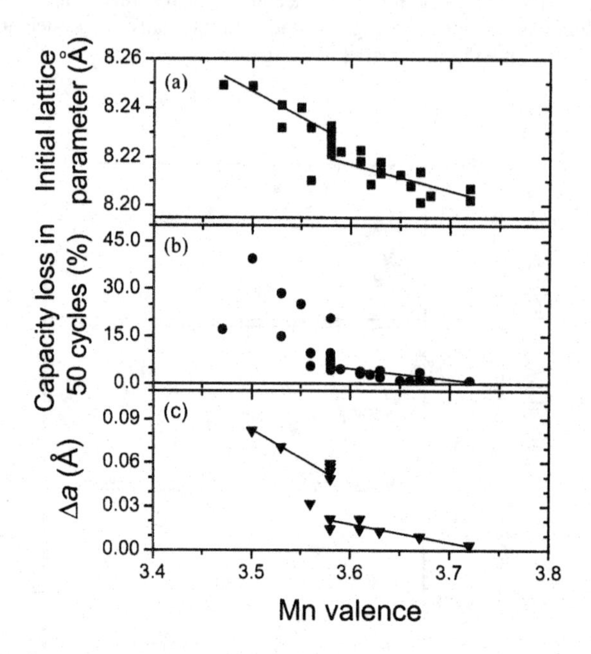

Figure 6: Correlation of the initial Mn valence of the as-prepared spinel oxides to the (a) initial lattice parameter of the as-prepared samples, (b) % capacity fade in 50 cycle, and (c) lattice parameter difference Δa between two cubic phases formed in the two-phase region during the charge-discharge process.

Figure 5c relates the lattice parameter difference Δa between the two cubic phases to the % capacity fade for several samples investigated in this study. The capacity fade decreases as the Δa value decreases. The lattice parameter difference can cause an instantaneous volume change in the oxide lattice, and a large instantaneous volume change (3 % in $LiMn_2O_4$) on going from one cubic phase to another cubic phase can lead to a loss of inter-particle contact and a breaking of particles during the discharge-charge cycling, resulting in poor Li^+ diffusion and electrical conductivity, unwanted side reactions, and capacity fade. It can be considered to be analogous to the instantaneous volume change (5.6 %) occurring during Jahn-Teller distortion in the 3 V region of $LiMn_2O_4$. In addition, as seen in Figure 6c, the difference in lattice parameters between the two cubic phases drop sharply around an initial Mn valance of 3.58+, resulting in a correlation between initial Mn valence and capacity fade.

CONCLUSIONS

Enhancement in the performance factors of batteries is essential to make them viable for electric and hybrid vehicles. Lithium ion batteries are attractive, but several issues need to be addressed to adopt them successfully for transportation applications. In this regard, the doubly substituted spinel lithium manganese oxides can become a good solution. For example, the doubly substituted spinel oxides, $LiMn_{1.85}Ni_{0.075}Li_{0.075}O_4$, show excellent cyclability at both room and elevated temperatures, higher rate capability than even $LiCoO_2$, good storage performance over the entire DOD range, little or small irreversible capacity loss in the first cycle, and low electrical resistance. The excellent electrochemical performance coupled with the inexpensive, abundant, and environmentally benign nature of Mn make the doubly substituted spinel manganese oxides appealing for transportation applications. The excellent performance of the doubly substituted spinel oxides compared to the unsubstituted $LiMn_2O_4$ appears to be related to the suppression of the lattice parameter difference Δa between the two cubic phases formed during the discharge-charge process and the consequent low lattice strain. Additionally, the electrochemical properties are found to bear a relationship to the initial lattice parameter and Mn valence. Samples with lower initial lattice parameters and a manganese valence of $> 3.58+$ show better performance as they lead to a smaller Δa.

ACKNOWLEDGEMENT

Financial support by the NASA Glenn Research Center and the Welch Foundation Grant F-1254 is gratefully acknowledged.

REFERENCES

[1]T. Q. Duong, J. A. Barnes, V. Battaglia, G. Henriksen, F. McLarnon, B.J. Kumar, and I. Weinstock, "U.S. Department of Energy Collaborative R&D on Electric and Hybrid Electric Vehicle Energy Storage Technologies: Current Status and Future Directions," 7B-1, The 20th International Electric Vehicle Symposium and Exposition, Long Beach, CA, Nov. 15-19, 2003.

[2]G. Amatucci and J. M. Tarascon, "Optimization of Insertion Compounds Such as $LiMn_2O_4$ for Li-Ion Batteries," *J. Electrochem. Soc.*, **149**, K31 (2002).

[3]H. Huang, C. A. Vincent, and P. G. Bruce, "Correlating Capacity Loss of Stoichiometric and Nonstoichiometric Lithium Manganese Oxide Spinel Electrodes with Their Structural Integrity," *J. Electrochem. Soc.*, **146**, 3649 (1999).

[4]Y. Shin and A. Manthiram, "Microstrain and Capacity Fade in Spinel Manganese Oxides," *Electrochem. Solid-State Lett.*, **5**, A55 (2002).

[5]J. H. Lee, J. K. Hong, D. H. Jang, Y. K. Sun, and S. M. Oh, "Degradation Mechanisms in Doped Spinels of $LiM_{0.05}Mn_{1.95}O_4$ (M=Li, B, Al, Co, and Ni) for Li Secondary Batteries," *J. Power Sources*, **89**, 7 (2000).

[6]D. H. Jang, Y. J. Shin, and S. M. Oh, "Dissolution of Spinel Oxides and Capacity Losses in 4 V $Li/Li_xMn_2O_4$ Cells," *J. Electrochem. Soc.*, **143**, 2204 (1996).

[7]M. M. Thackeray, Y. Shao-Horn, A. J. Kahaian, K. D. Kepler, E. Skinner, J. T. Vaughey, and S. A. Hackney, "Structural Fatigue in Spinel Electrodes in High Voltage (4 V) $Li/Li_xMn_2O_4$ Cells," *Electrochem. Solid-State Lett.*, **1**, 7 (1998).

[8]R. V. Chebiam, F. Prado, and A. Manthiram, "Soft Chemistry Synthesis and Characterization of Layered $Li_{1-x}Ni_{1-y}Co_yO_{2-\delta}$ ($0 \leq x \leq 1$ and $0 \leq y \leq 1$)," *Chem. Mater.*, **13**, 2951 (2001).

[9]R. A. Young, A. Shakthivel, T. S. Moss, and C. O. Paiva-Santos, "DBWS-9411 - an Upgrade of the DBWS*.* Programs for Rietveld Refinement with PC and Mainframe Computers," *J. Appl. Crystallogr.*, **28**, 366 (1995).

[10]Y. Shin and A. Manthiram, "Influence of the Lattice Parameter Difference between the Two Cubic Phases Formed in the 4 V Region on the Capacity Fading of Spinel Manganese Oxides," *Chem. Mater.*, **15**, 2954 (2003).

HRTEM IMAGING AND EELS SPECTROSCOPY OF LITHIATION PROCESS IN FEFX:C NANOCOMPOSITES

F. Cosandey, J.F. Al-Sharab, F. Badway and G.G. Amatucci
Department of Ceramic and Materials Engineering
Rutgers University
Piscataway, NJ 08854-8065

ABSTRACT

A new type of positive electrodes for Li-Ion batteries has been developed based on FeF_3:C carbon metal fluoride nanocomposites. The electrochemical data reveal a reversible metal fluoride conversion process with a high specific capacity of about 600 mAh/g realized at 70°C. The nanocomposite electrodes were analyzed after synthesis as well as upon voltage cycling from 4.5 to 1.5V by combined selected area electron diffraction (SAED), high-resolution transmission electron microscopy (HRTEM) and electron energy-loss spectroscopy (EELS) techniques. HRTEM combined with SAED were used for phase identification while EELS was used to determine the valence state of Fe from L_3 energy shift as well as white line L_3/L_2 intensity ratio. After synthesis, the microstructure is composed of FeF_3 nano-crystal of the order of 15 nm encompassed in a matrix of carbon. Upon discharged to 1.5V, a complete reduction of iron to Fe^o is observed with the formation of a finer Fe:LiF:C nanocomposites (~8 nm). Upon recharging to 4.5V, EELS and HRTEM data reveal a reoxidation process to a Fe^{2+} state with the formation of a carbon metal fluoride nanocomposite related to the FeF_2 structure.

INTRODUCTION

At the present time, the Li-Ion battery technology is dominated by the layered Li intercalated compound $LiCoO_2$ with more recent developments based on layered manganate ($LiMnO_2$) and phosphate materials ($LiMePO_4$ and $LiMe(PO_4)_3$ where Me is a transition metal (Cr, Mn, Fe, Co, Ni...). The fundamental route to obtain the highest capacity batteries is to utilize all the possible oxidation states of the compound electrodes during charge and discharge cycles. For the $LiCoO_2$ standard electrode, only one electron per Co ion is utilized during the complete electrochemical conversion to CoO_2. Furthermore, for $LiCoO_2$ and in most electrode materials, a complete redox reaction is not achieved reversibly limiting the capacity even further. Recently, we have synthesized a new type of positive electrodes for Li-Ion batteries based on FeF_3:C nanocomposites [1-3] were the metal ion has an initial 3+ valence states and were a complete reversible redox reaction is expected according to the following electrochemical reaction.

$$3Li^{+} + 3e- + Fe^{3+}F_3 = 3LiF + Fe^{o} \qquad (1)$$

We have studied metal fluorides nanocomposites (CMFNC) at room temperature and at 70°C under identical discharged-charged rates of 7.58 mA/g [3]. For the second cycle voltage curve i.e. after re-charging, it was observed that the FeF_3:C nanocomposites has a specific capacity at 70°C of 620 mAh/g. Such a high capacity is indicative of a $3e^-$ transfer as represented by Equ.1. However, recent crystallographic data obtained by SAED points towards a re-conversion to FeF_2 involving only $2e^-$ transfer [3]. In order to better understand the behavior of these new metal-fluoride electrode material during the charge and discharge reactions it is necessary to determine the structural evolutions that are occurring and in particular to determine the valence state of Fe. For transition metals (Fe, Mn, Co,...), their valence state can be determined uniquely by electron energy loss spectroscopy (EELS) from the shift in L_3 line energy and from the L_3/L_2 line intensity ratio [4-6].

In this paper, we report results on the microstructural evolution during charge-discharge processes of litiation of FeF_3:C nanocomposites by combined HRTEM and EELS.

EXPERIMENTAL PROCEDURES

The carbon metal fluoride nanocomposites (CMFNCs) were fabricated in He by the high-energy ball milling of FeF_3 with activated carbon. Stoichiometric mixtures were placed inside a hardened steel mill cell along with hardened steel media. Unless otherwise noted, 85:15 % FeF_3:C were fabricated by high energy milling for 60 min. For electrochemical characterization, electrodes were prepared by adding poly(vinylidene fluoride-co-hexafluoropropylene), carbon black and dibutyl phthalate to the CMFNC in acetone. The slurry was tape cast, dried for 1 hour at 22°C and rinsed in 99.8% anhydrous ether to extract the dibutyl phthalate plasticizer. The electrodes, $1cm^2$ disks typically containing 50±1% CMFNC and 18±1% carbon black were tested electrochemically versus Li metal. Swagelock or coin cells were assembled in a He-filled dry box using glass fiber separators saturated with 1M $LiPF_6$ in ethyl carbonate: dimethyl carbonate (EC: DMC 1:1 in vol.) electrolyte. The electrochemical cells were cycled under constant current of 7.58 mA/g and were controlled by Mac-Pile or Maccor battery cycling systems. Further details of nanocomposite synthesis and electrochemical measurements have been published elsewhere [2].

Ex-situ TEM of the positive nanocomposite electrode during various stages of discharge-charge was accomplished by first disassembling the electrochemical cells containing powder electrodes. The powders were then rinsed in dimethyl carbonate and samples for TEM observation were prepared by dispersing the powders in trichlorotrifluoroethane and releasing a few drops of the liquid on a "lacey" carbon film supported on a copper mesh. All preparations were accomplished in an anhydrous atmosphere. The grids were then sealed under helium and transferred to the TEM room. The final transfer to the microscope consisted of less than 30s exposure to ambient atmosphere. Standard dark field, selected area electron diffraction (SAED) as well as high-resolution lattice images (HRTEM) were obtained using a Topcon 002B microscope operating at 200kV. HRTEM images were also obtained with a FEI CM-300 FEG operating at 300 KV. The EELS spectra were obtained with a Gatan 666 PEELS spectrometer attached to a Hitachi HF2000. Details in EELS spectra acquisition and data processing can be found in a companion paper [7].

RESULTS AND DISCUSSION

Microstructural characterization of FeF_3:C nanocomposites.

A typical HRTEM image of as synthesized FeF_3:C nanocomposite obtained by high-energy milling coarse FeF_3 powders with activated C is shown in Fig.1a with in Fig.1.b the particle size distribution measured from dark-field TEM images.

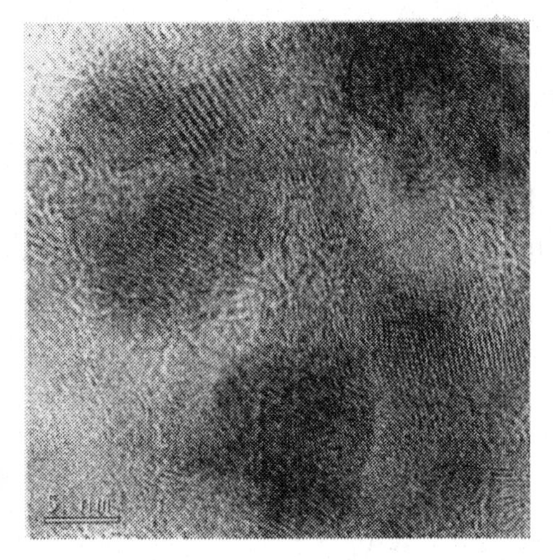

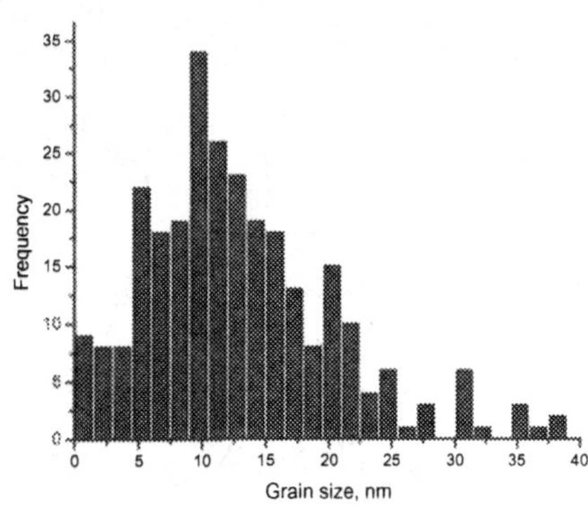

Figure 1. (a) High-resolution TEM image of FeF_3/C nanocomposite, (b) particle size distribution obtained from dark-field TEM image

From Fig.1, it can be seen that the microstructure is composed of nano-sized FeF_3 crystals embedded in a carbon matrix. The average size is measured to be 15 nm. Lattice spacings obtained from FFT diffractogram reveal d spacings corresponding to the FeF_3 structure (FeF_3 is trigonal (R-3c) with a=0.514 nm and c= 1.41nm) with a few lattice spacings corresponding to FeF_2 (FeF_2 is tetragonal (P42mnm) with a=0.469 nm and c=0.333 nm). This result is in agreement with x-ray and selected area electron diffraction (SAED) data revealing also the appearance of a minor FeF_2 phase after prolonged milling times [3].

Electrochemical characterization of FeF_3:C nanocomposites.

The FeF_3:C nanocomposites were characterized electrochemically versus a Li metal counter electrode in the range of 4.5 to 2 V at constant current of 7.58 mA/g. The effect of initial FeF_3 particle size on cell performance is shown in Fig.2. For manual mixing with a particle size of 120 nm, the performance is poor at 50 mAh/g and remains stable only for a small period of time. After milling for 240 min the particle size deceases to 15 nm with a drastic increase in capacity to 210 mA/g with respectable stability. In an earlier study on iron trifluoride (FeF_3) only a limited capacity of 80mAh/g was reported for a discharge voltage region from about 4.5V to 2V at an approximate rate of C/20 [8]. The extreme ionicity of the fluorides resulting in very poor electronic conductivity combined with a questionable ionic conductivity leads to the disparity between 80mAh/g and the theoretical (1e$^-$ transfer) capacity of 237mAh/g. The present results shown in Fig.2 clearly indicate that in FeF_3:C nanocomposite with particle sizes below 20 nm, these intrinsic conductivity limitations are overcome.

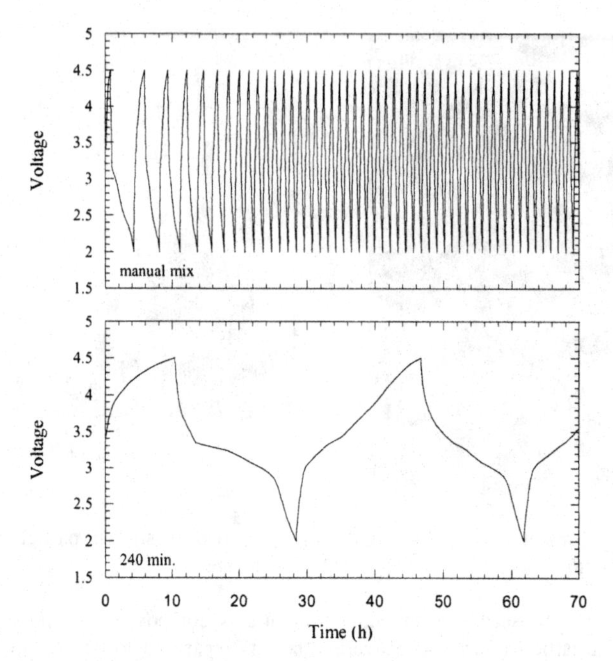

Fig. 2. Voltage versus time plot for Li-FeF_3:C cells cycled at 7.58 mA/g of composite after manual mixing and after milling for 240 min., leading to particle size of 120 nm and 15 nm respectively.

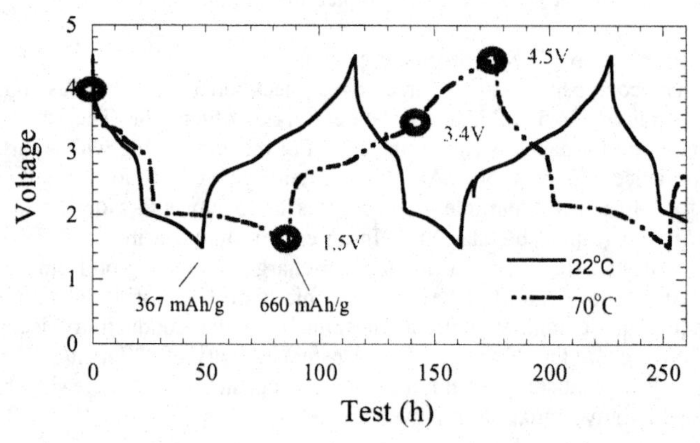

Fig. 3. Comparison of discharge/ charge voltage profile of FeF_3:C nanocomposite at 22 °C and 70°C and at constant current of 7.58 mA/g. Data shows first and second full discharged after a short initial charge to 4.5 V

The effect of cell temperature is shown in Fig.3 for two temperatures of 22 °C and 70 °C. The reversible electrochemical reaction is kinetically limited with a higher capacity of 660 mAh/g measured at the higher operating temperature. This specific capacity is close to the theoretical capacity (3x273 mAh/g) for the reversible reaction involving three-electron transfer. These capacities are more than twice the current capacity of the commercial $LiCoO_2$ electrode materials.

Microstructural characterization of discharged and charged electrodes

In order to follow microstructural changes the during charge and discharge processes, HRTEM, SAED and EELS was used to determine the structure and valence state of the nanocomposite after synthesis and after full discharged at 1.5V and re-charged states of 3.4 and 4.5 V as shown by circles in Fig.3.

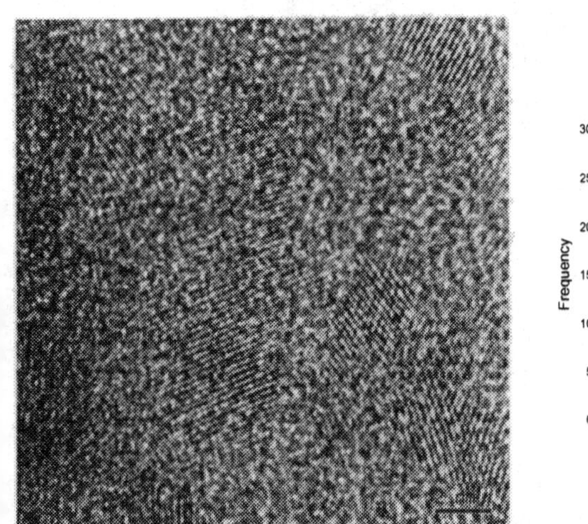

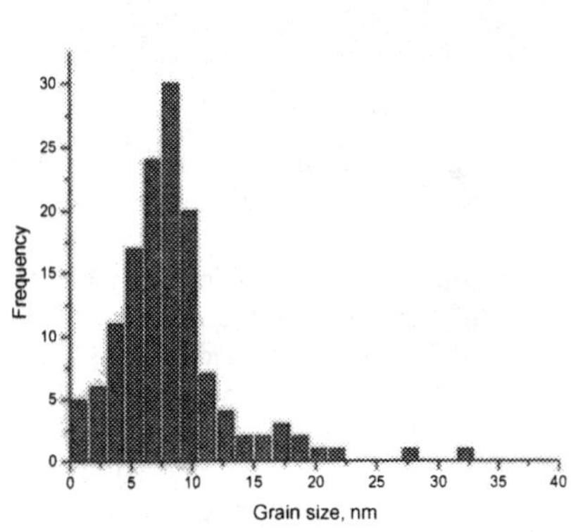

Figure 4. (a) High-resolution TEM image of FeF_3:C nanocomposite electrode discharged to 1.5V and (b) particle size distribution obtained from dark-field TEM image

A HRTEM image of discharged electrode to 1.5V is shown in Fig. 4a with in Fig. 4b, the particle size distribution measured from dark field TEM images. Lattice spacing measurements reveal that the structure consists of Fe and LiF consistent with SAED analysis [3]. During the conversion process, the particle size reduces to an average size of 8nm from an as synthesized size of 15 nm. Upon re-charging to 4.5 V, the particle size reduces even further and is now of the order of 6-7 nm as shown in Fig. 5b. Lattice spacing measurements taken from FFT of HRTEM image of Fig. 5a reveal lattice spacings for FeF_2 type structure.

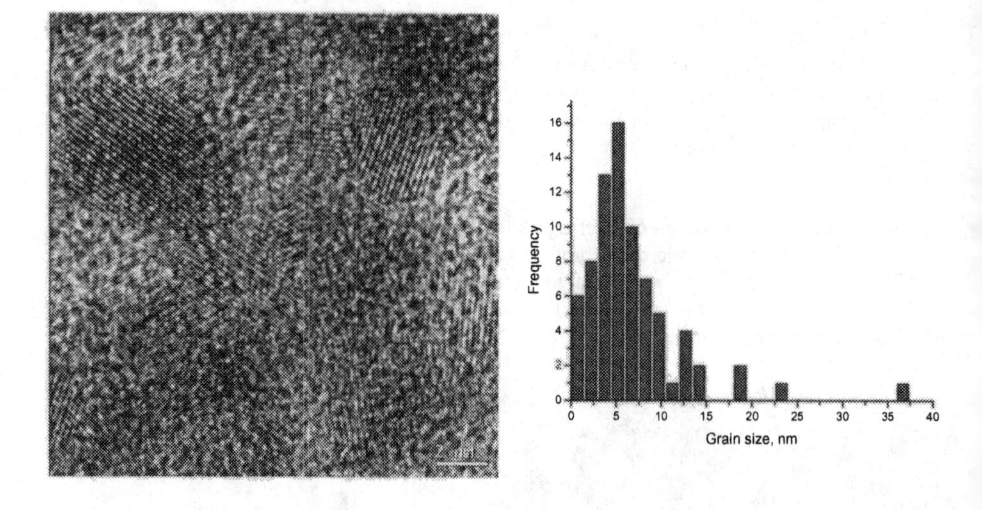

Figure 5. (a) High-resolution TEM image of FeF_3:C nanocomposite electrode re-charged to 4.5V and (b) particle size distribution obtained from dark-field TEM image.

Valence state determination of Fe by EELS

In order to follow quantitatively the chemical state of Fe during charge and discharge processes, we have examined EELS spectra of the standard LiF, Fe, FeF_2 and FeF_3 compounds, the as synthesized FeF_3:C nanocomposite as well as electrode material taken at various stages of lithiation corresponding to charge voltages of 1.5, 3.4 and 4.5 V. Typical EELS spectra for FeF_2 and FeF_3 are shown in Figure 6 altogether with re-charged electrode to 4.5 V. The L edge of FeF_2 and FeF_3 is characterized by two L_3 and L_2 lines resulting from the transition of electrons from the spin-orbit split levels $2p_{3/2}$ and $2p_{1/2}$ to unoccupied 3d states. The other peak observed at 695 eV corresponds to the K edge of F resulting from a transition of electrons from the 1s level to unoccupied 2p states. For FeF_3 there is an additional pre-peak marked P located at 684 eV. The origin of this pre-peak can be attributed to a forbidden 1s to 3d transition indicating that the unoccupied 2p states hybridized with 3d Fe orbital giving rise to a measurable covalent effects in FeF_3 [9].

For all samples, we have measured the energy of the peak L_3 as well as the L_3/L_2 intensity ratio. Precise L_3 energy measurements were obtained by first calibrating the spectrometer energy with respect to the zero loss that was further refined by normalization with respect to C k edge energy from the support film. The L_3/L_2 intensity ratio we obtained by a series of steps involving the removal of the background, the contribution form F signal of LiF and the continuous L edge from Fe. The integrated L_3 and L_2 line intensities were calculated using a 4 eV window centered on the L line maxima. Additional analysis details can be found elsewhere [7]. The results of this analysis are summarized in Table II. From this Table II, it can be seen that the L_3 peak energy increases from 708.9 eV to 710.3 eV for Fe^{2+} and Fe^{3+} respectively. The L_3 peak energy of Fe^{o} has a similar value as Fe^{2+}. On the other hand, the L_3/L_2 intensity ratio increases from a low value of 2.9 for Fe^{o} to 4.4 and 4.2 for Fe^{2+} and Fe^{3+} respectively. For iron oxides, it has been shown that the L_2/L_3 intensity ratio increases with increasing valence [5] and by comparison, this ratio should be larger for FeF_3 than for FeF_2. This is in contrast to our experimental values. However, there is some error associated with our ratio measurements due to the noise of the signal and the removal procedure for the F edge.

Upon inspection of Table II it can be seen that for the 1.5 V sample, the Fe valence correspond to metallic Fe^o from the low value of the L_3/L_2 ratio. Upon increasing voltage to 3.4V, the L_3/L_2 ratio increases indicative of an increase in valence state for Fe. This ratio is however lower than the value for Fe^{2+} indicating that the transformation is not complete or homogeneous and that there is some residual Fe^o left in the cathode material. After full charging to 4.5V, the L_3/L_2 ratio increases further to a value corresponding to Fe^{2+} or Fe^{3+}. From the associated lower L_3 peak energy of 708.4 V, we can conclude that iron has definitely the Fe^{2+} valence state at 4.5V. This can be seen clearly in Fig.6 from the similitude between spectra for FeF_2 and 4.5 V, and when compared to FeF_3, from its lower L_3 peak energy and the absence of the pre-peak marked P.

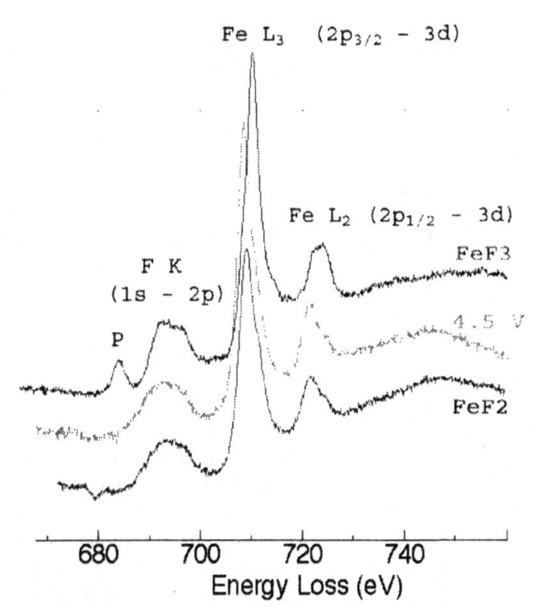

Figure 6. EELS spectra for the two standard fluoride compounds (FeF_2 and FeF_3) with the EELS spectra for FeF_3/C nanostructured electrode re-charged to 4.5V

These EELS results confirms unequivocally earlier findings that the reconverted phase is related in structure to $Fe^{2+}F_2$ not $Fe^{3+}F_3$ as anticipated from Equ.1 [3]. This result is surprising in view of the reversible capacity of 660 mAh/g reflecting a $3e^-$ transfer in the reversible electrochemistry as opposed to $2e^-$ expected for a FeF_2 phase conversion. As presented earlier in Fig. 5, the reversible electrochemical reduction and reconversion processes leads to a FeF_2:C nanocomposite with an average particle size of 6-7 nm. With such a large surface area in contact with the electrolyte, it is possible to speculate that a large fraction of the grains have active interfaces which could enable the oxidation of $Fe^{2+}F_2$ to $Fe^{3+}F_3$ by a pseudo-capacitive reaction. This reaction would involve the surface adsorption of PF^-_6 to form a surface entity $Fe^{3+}F_2(PF_6^-)$ upon the anodic step.

Table II. Fe L_3/L_2 line intensity ratio and Fe L_3 peak energy (eV) for FeF_3:C nanocomposite, the discharged 1.5 V, partially charged 3.4 V and fully charged 4.5 V positive electrode material and for the three Fe valence states of the standard Fe, FeF_3, and FeF_2 compounds.

Sample	Fe L_3/L_2 Line Intensity Ratio	Fe L_3 Peak Energy (eV)
$Fe^{3+}F_3$	4.2	710.3
$Fe^{2+}F_2$	4.4	708.9
Fe^0	2.9	708.5
$Fe^{3+}F_3$:C	-	711.0
1.5 V	2.8	708.8
3.4 V	3.6	708.6
4.5 V	4.3	708.4

SUMMARY-CONCLUSIONS

A new type of positive electrodes for Li-Ion batteries has been developed based on FeF_3:C carbon metal fluoride nanocomposites. Electrochemical data reveal a metal fluoride conversion process involving $3e^-$ with a high specific capacity of about 600 mAh/g realized at 70°C. HRTEM combined with SAED were used for particle size measurements while EELS was used to determine the valence state of Fe. Upon discharged to 1.5V, a complete reduction of iron to Fe^O is observed with the formation of a fine Fe:LiF:C nanocomposites (~8 nm). Upon recharging to 4.5V, EELS reveal a reoxidation process to a Fe^{2+} state with the formation of a carbon metal fluoride nanocomposite 6-7nm in size and with the FeF_2 structure.

ACKNOWLEDGEMENT

This work is supported by the US government. Special thanks for P. Stadelmann of CIME-EPFL, Switzerland, for numerous discussion and for the use of the electron microscopy facility.

REFERENCES

1. I. Plitz, F. Badway, J. Al-Sharab, A. DuPasquier, F. Cosandey, and G.G. Amatucci, *Carbon metal fluoride nanocomposites: Structure and electrochemistry of carbon-metal fluoride nanocomposites fabricated by a solid state redox conversion reaction.* [Submitted for publication] (2004)
2. F. Badway, N. Pereira, F. Cosandey, and G.G. Amatucci, *Carbon metal fluoride nanocomposites: Structure and electrochemistry of FeF3:C.* J. of Electrochem. Soc., **150** [9] A1209-A18 (2003)
3. F. Badway, F. Cosandey, N. Pereira, and G.G. Amatucci, *Carbon metal fluoride nanocomposites: High capacity reversible metal fluoride conversion materials as rechargeable positive electrodes for Li batteries.* J. Electrochem. Soc., **150** [10] A1209-A18 (2003)
4. G.A. Botton, C.C. Appell, A. Horsewell, and W.M. Stobbs, *Quantification of the EELS near-edge structure to study Mn doping in oxides.* Journal of Microscopy, **180** [3] 211-16 (1995)
5. C. Colliex, T. Manoubi, and C. Ortiz, *Electron-energy-loss-spectroscopy near-edge fine stuctures in the iron-oxygen system.* Physical Review, **B44** [20] 11403-11 (1991)
6. D.H. Pearson, B. Fultz, and C.C. Ahn, *Measurements of 3d state occupancy in transition metals using electron energy loss spectrometry.* Appl. Phys. Lett., **53** [15] 1405-07 (1988)
7. F. Cosandey, J. Al-Sharab, F. Badway, G.G. Amatucci, and P. Stadelmann, *EELS spectroscopy of FeFx/C nanocomposites.* (to be published), (2004)
8. H. Arai, S. Okada, Y. Sakurai, and J. Yamaki, *Cathode performance and voltage estimation of metal trihalides.* J. Power Sources, **68** 716-19 (1997)
9. M.S.M. Saifullah, G.A. Botton, C.B. Boothroyd, and C.J. Humphreys, *Electron energy loss spectroscopy studies of the amorphous to crystalline transition in* FeF_3. J. Applied Physics, **86** [5] 2499-504 (1999)

AMORPHOUS SILICON THIN FILM ANODES FOR LITHIUM-ION BATTERIES

J. P. Maranchi and P.N. Kumta
Carnegie Mellon University
Dept. of Materials Science and Engineering
Pittsburgh, PA 15213

A.F. Hepp
NASA Glenn Research Center
PV and Space Environments Branch
Cleveland, OH 44135

ABSTRACT

Due to its order of magnitude higher theoretical capacity, silicon has been widely proposed as a potential next-generation anode material to replace conventional carbon based lithium-ion battery anodes. In this study, the properties of amorphous 250 nm and 1 μm silicon films deposited by R.F. magnetron sputtering on copper foil are investigated using x-ray diffraction, electron microscopy, and electrochemical methods. Battery tests have shown that the 250 nm Si thin films exhibit an excellent reversible specific capacity of nearly 3500 mAh/g when tested for 30 cycles. The high reversible capacity and excellent cyclability of the 250 nm sputtered silicon thin films suggest excellent adhesion between Si and Cu leading to high capacity retention. SEM analysis conducted on the 250 nm Si films after the 30th charge suggests the good adhesion of the ~ 2 μm diameter "plates" of silicon to the copper substrate.

INTRODUCTION

Despite the widespread commercialization of lithium-ion batteries employing carbon based anode materials, intense research is currently ongoing to identify higher capacity anode materials for the next generation of lightweight and compact lithium-ion batteries. Original research conducted on bulk alloy type anodes comprising elemental tin or silicon showed their potential for high capacity. Contrary to intercalation type electrodes, the alloy type anodes experience severe volume expansion and contraction resulting in crumbling and mechanical degradation.[1] In an effort to reduce the harmful effects of decrepitation, multi-component tin containing oxides were used which yielded substantial improvements in cyclability.[2] However, the based oxide materials exhibit irreversible capacities as high as ~50% of the total capacity.[3] In order to reduce the irreversible capacity, researchers have used non-oxide based inactive materials to withstand the volumetric strains of the active phases. Metallic alloys containing active and inactive phases have also showed promise.[1,3-6] Other researchers recognized the importance of small grain size along with a ductile inactive phase in order to stabilize the capacity of alloy type anodes.[7] Pyrolized polymers were also used to produce nano-dispersed silicon in a carbon matrix yielding reversible specific capacities up to 600 mAh/g.[8-10] Research in our group has demonstrated that the amorphous phase of silicon can be contained within nanocrystalline matrices of TiN, TiB_2, and SiC to successfully stabilize reversible capacities up to 400 mAh/g.[11-13] Our group has also shown that carbon can be used as a matrix material in both Sn and Si anode systems. Recently, Song *et al* have reported capacities as high as ~2200 mAh/g in nanocrystalline thin films of Mg_2Si 30 nm thick deposited by pulsed laser deposition.[16] Although these systems yield specific capacities greater than graphite (372 mAh/g), there have been very few reports to date on the electrochemical response of materials or systems exhibiting near theoretical specific capacity of silicon (~4200 mAh/g).[17,18]

Very few researchers have investigated thin or thick films of silicon or silicon-tin. Bourderau *et al* had limited success studying CVD deposited thick (1.2 μm) amorphous silicon films on porous nickel substrates.[19] More recently, Lee *et al* studied the cycle related stress effects in sputtered silicon thin films.[20] Reversible, large scale volume changes were seen in meso-scale sputtered Si-Sn films reported by Beaulieu *et al.*[21] Research by Beaulieu *et al* provided good insight into the nature of the electrochemical reaction of lithium with sputtered amorphous $Si_{0.66}Sn_{0.34}$ films.[22] Several groups have also presented encouraging results on Si films deposited on Cu and Ni substrates at the 11th International Meeting on Lithium Batteries (IMLB).[23-25]

The promising results of sputtered films prompted the use of sputtering as a synthetic tool in this study to research the electrochemical behavior of amorphous silicon on an inactive, ductile copper foil substrate. Preliminary cycling data show promising results for both the 1 μm and 250 nm Si films with the 1 μm films exhibiting reversible capacities of 3000 mAh/g for 12 cycles, while the 250 nm films exhibit reversible capacities of ~3500 mAh/g for 30 cycles with no obvious signs of failure. This paper reports the experimental studies conducted on amorphous Si films deposited on Cu. The results show promising reversibility in an amorphous Si anode system.

EXPERIMENTAL

Nanoscale and mesoscale films of Si were prepared by r.f. magnetron sputtering (Perkin Elmer – 8L) from a commercial 5" diameter Si target (99.999%) at 200 W onto high purity oxygen free (HOFC) 0.001" thick Cu foil substrates (Insulectro). The base pressure was $5x10^{-7}$ Torr and the working pressure was 5 mTorr argon. Pre-sputtering was done for 30 min. at 400 W prior to each deposition to remove any surface oxides from the target. 250 nm and 1.0 μm films were deposited at a rate of ~ 86 Å/min. Prior to depositing the Si films on Cu foil, the sputtering rate was determined in a rate deposition experiment using the same experimental parameters desired for the actual deposition run (conducted on smooth Corning 0211 glass substrates partially masked by Kapton tape.) The thickness was determined by averaging seven step-heights measured using a profilometer (Tencor). From the deposition time and thickness, an accurate sputtering rate was determined. The accuracy of the profilometer thickness measurements was confirmed on the Cu foil - 250 nm Si films by cross-sectional transmission electron microscopy (TEM, Jeol Electron Microscopy-2000EX II, 200 kV, length calibration performed using a standard replica Cu calibration grid). All substrates were cleaned with acetone and alcohol followed by drying with compressed nitrogen. Following deposition, the substrates were allowed to cool for 30 min. on a water-cooled platen before removal from the vacuum chamber.

The phases present in the as-deposited films were analyzed by X-ray diffraction incorporating a state-of-the-art detector (XRD, Philips PW3040PRO, θ/θ powder diffractometer with X'celerator detector and Cu radiation source), and the microstructure and chemical composition of the films was examined using a scanning electron microscope (SEM, Philips XL30 equipped with energy dispersive X-ray analysis, EDAX). The film-substrate interface was examined in detail using a high-resolution transmission electron microscope (Philips Tecnai F20 FEGTEM, 200 kV).

The electrochemical properties of the sputtered Si samples were examined using 2016 coin cells (Hohsen). 1 cm^2 disks were punched from the Cu foil/Si film samples and placed inside an argon filled glove box (Vacuum Atmospheres, Hawthorn, CA, < 10 ppm O_2 and H_2O) for fabrication of half-cells. The half-cells were fabricated in the glove box using lithium foil and a micro-porous polyethylene separator (Tonen) together with the deposited films and 1 M $LiPF_6$ in ethylene carbonate/dimethyl carbonate (2:1) electrolyte (Merck). The crimped coin cells were removed from the glove box and allowed to equilibrate for 12 hours prior to galvanostatic cycling (Arbin Instruments). All charge/discharge experiments were 30 cycles in duration and the cycling voltages were limited between 20 mV and 1.2V. Three Si films of both thicknesses were tested employing a C/2.5 current rate while two films of both thicknesses were cycled at 2C rate (based on theoretical capacity of silicon) using a 60 s rest period between each charge/discharge cycle.

RESULTS AND DISCUSSION

The XRD results of the Cu substrate and as-deposited films are shown in Fig. 1. The only peak attributed to the silicon films is a low intensity, very broad peak centered near 29° two-theta. All of the other peaks are due to the Cu foil substrate. Although a Ni filter was used in the diffracted beam optics, both Kα and Kβ peaks are seen for the Cu substrate due to the long integration time. The XRD results suggest that the as-deposited Si films are amorphous. It is believed that the amorphous starting materials lead to enhanced cyclability and cause homogeneous volume expansion/contraction by eliminating the existence of any two phase regions during charge/discharge of the material.[21] In order to conclusively determine the short or medium range atomic arrangement in the as-prepared Si films, more detailed characterization using transmission electron microscopy (TEM) will be necessary.

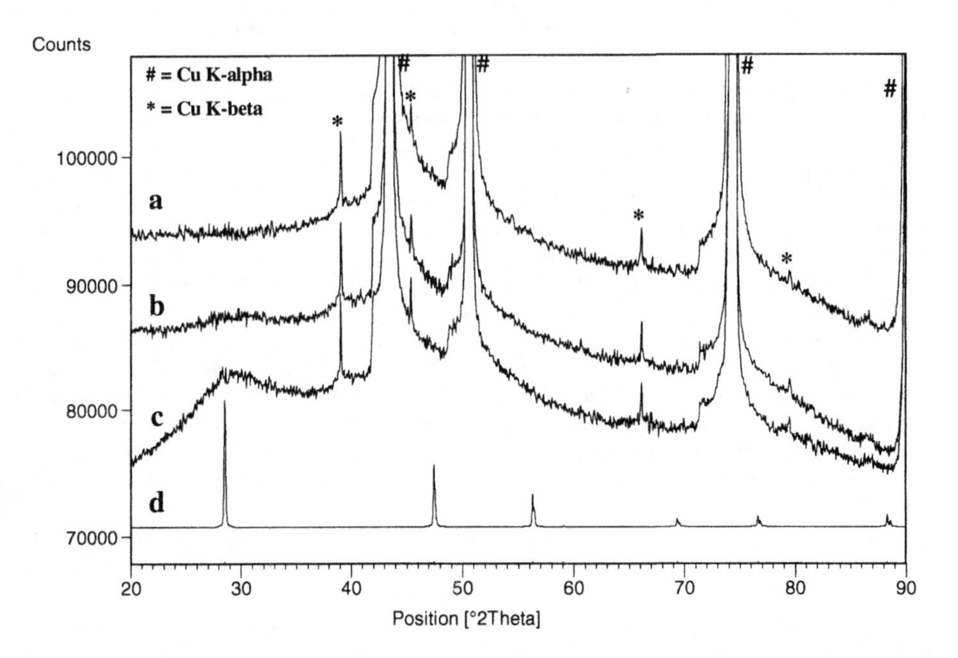

Figure 1. XRD patterns of (a) Cu foil substrate only, (b) 250 nm Si on Cu foil. (c) 1.0 μm Si on Cu foil, (d) simulated pattern of polycrystalline Si using data from JCPDS database[26].

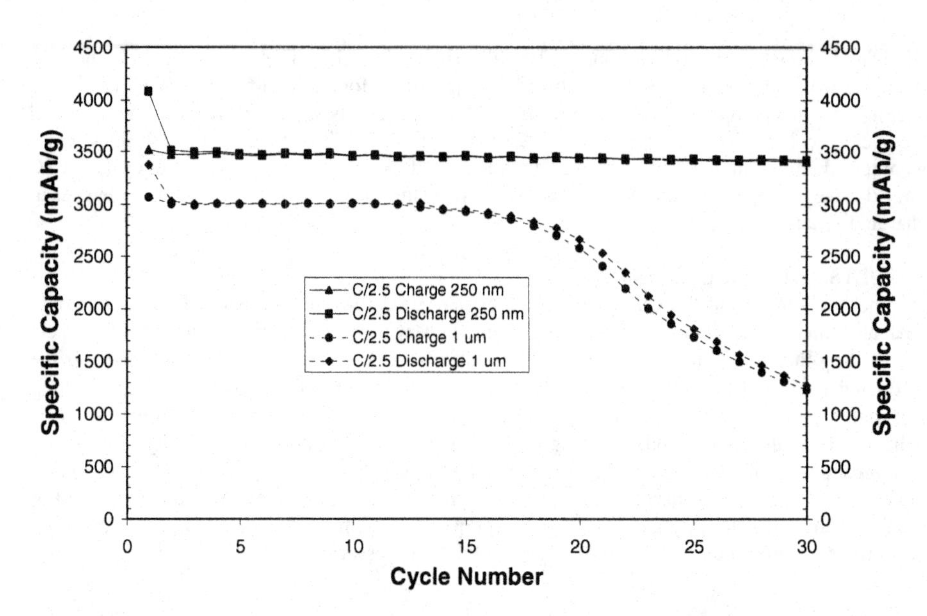

Figure 2. Comparison of galvanostatic cycling of 250 nm and 1.0 μm Si films on Cu foil (both at ~C/2.5 rate).

The electrochemical half-cell cycling results of the Si films on Cu foil show that the 250 nm films exhibit a far superior cyclability and have higher specific capacity than the 1.0 μm thick films. The effect of thickness on the specific capacity and cyclability of the Si films on copper is displayed in Fig. 2. The thickness of the 250 nm Si film was verified by cross-sectional TEM analysis, as shown in Fig. 3a. The gravimetric capacity was calculated using the thickness and area of the films, assuming an estimated theoretical density of 2.33 g/cc for silicon. Use of the literature value of 2.33 g/cc for the density of silicon is justified because it is the highest possible density for the as-deposited films, giving the most conservative estimate for the specific capacity of the material. One can see that the thickness of the as-deposited films affects both the reversible capacity and cyclability. The 250 nm film exhibits a first cycle discharge capacity of nearly 4100 mAh/g and a reversible capacity of ~3500 mAh/g. On the other hand, the 1 μm film showed a first cycle discharge capacity of 3300 mAh/g and a reversible capacity of 3000 mAh/g. The increase in thickness from 250 nm to 1 μm however results in capacity fade after only 12 cycles while the 250 nm film remains stable up to 30 cycles. In both cases, the capacity observed is very impressive, suggesting the potential viability of these films as future anode materials.

One possible explanation for the excellent cyclability of the 250 nm thick Si film compared to the 1.0 μm film is the residual stress in the film. Physical observations of the post-deposition films showed that the Si coated Cu foils would bend and begin to curl (convex on the Si face), with the degree of bending and curling being much higher in the thicker film indicating a higher film stress. Future research on the role of film stress in the cyclability of the films will be performed. The film stress may also relate to the capacity fade seen in the 1.0 μm sample after 12 cycles. A possible reason for the reversible capacity of both films being less than the

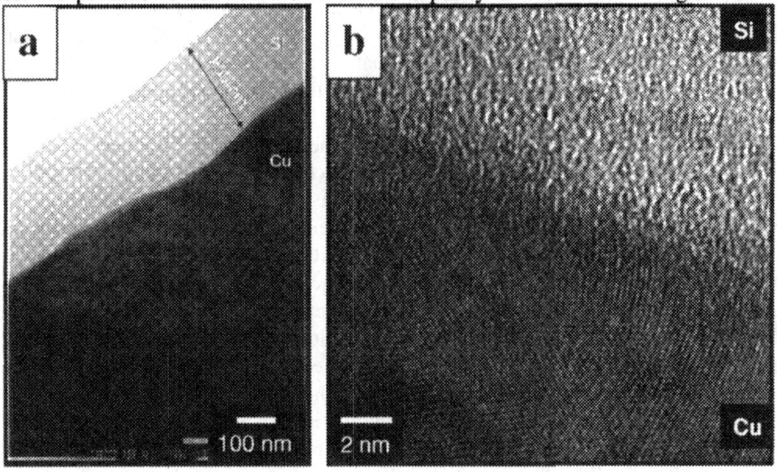

Figure 3. (a) Conventional cross-sectional TEM bright-field image of the as-deposited 250 nm Si film on Cu , and (b) same sample as shown in Fig. 3(a), cross-sectional HR-TEM image of the Cu-Si interface.

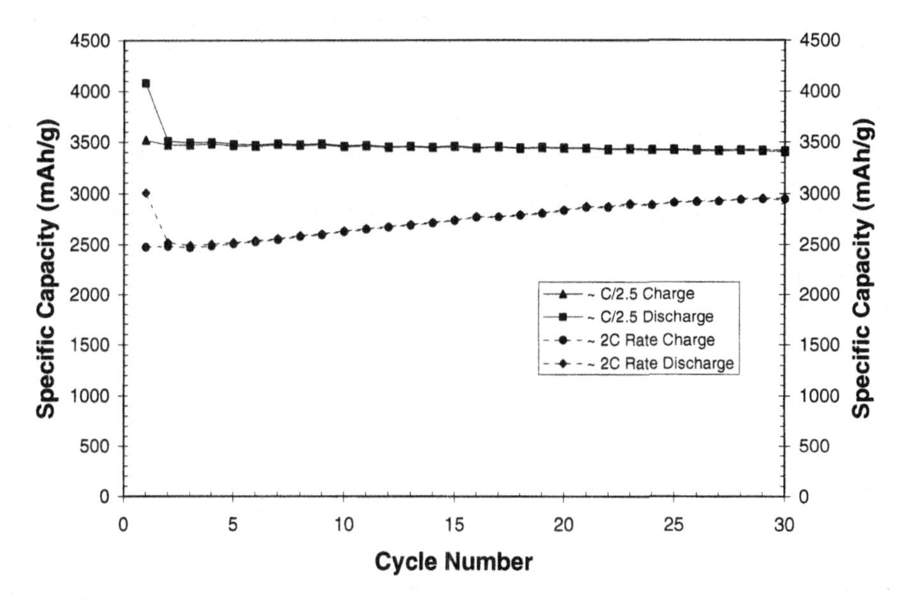

Figure 4. Comparison of galvanostatic cycling rate of 250 nm Si films on Cu foil (~C/2.5 and ~2C rates).

theoretical capacity of Si is the formation of an amorphous or near amorphous interfacial Cu-Si interfacial phase (during sputtering or electrochemical cycling). The formation of a Cu-Si interfacial layer will result in the loss of active silicon leading to lower capacity. High-resolution cross-section TEM investigation of the Cu-Si interface, as shown in Fig. 3b, shows no obvious interfacial phase. One possible reason can be suggested for the first cycle irreversible loss observed in both Si film thicknesses. De-lamination could occur during the first charge of a small portion of the film closest to the edge due to fracture induced in the film edge during electrode punching.

A preliminary study of the rate capability of the 250 nm Si film on Cu substrate was performed because of the excellent charge/discharge characteristics observed at C/2.5 rate. A previously uncycled film from the same sputtering run was tested for 30 cycles at 2C rate. The specific capacity versus cycle number of the films tested at 2C and C/2.5 rates are shown in Fig. 4 for comparison. The sample cycled at 2C rate exhibits a first cycle discharge of ~3000 mAh/g and an initial reversible capacity of ~2500 mAh/g. The specific capacity of the film cycled at a high rate of 2C increases with every cycle ultimately exhibiting a value of ~3000 mAh/g at the end of the 30th cycle with near perfect (~100%) coulombic efficiency. In the 30th cycle, the difference in specific capacity between the films cycled at C/2.5 and 2C rates is ~400 mAh/g. The lower capacity observed in the film cycled at a higher rate of 2C compared to that of the film tested at a C/2.5 rate could be due to the kinetic limitations of alloying of lithium with elemental silicon at room temperature. Future rate studies will focus on the electrical conductivity of the thin films, as well as the lithium-ion diffusivity at room temperature. A further study of the films cycled at 2C rate for longer cycles is warranted to observe the upper limit of the reversible specific capacity. Nevertheless, these preliminary experiments conducted on the rate capability of the 250 nm Si films on Cu indicate that the sputter-deposited silicon is a fairly rate tolerant anode material.

The voltage versus time profile is shown in Fig. 5(a). The voltage profile has a flat region in the first charge and a smoother, sloped shape in subsequent cycles with good reversibility after the first cycle up to thirty cycles. From the corresponding calculated differential capacity (dQ/dV) plot shown in Fig. 5(b), one can see some sharp peaks during the first discharge of the lithium alloying step, indicating the presence of two phase regions during the reaction of lithium with the silicon film. The two phases leading to the two sharp peaks in the first discharge of the

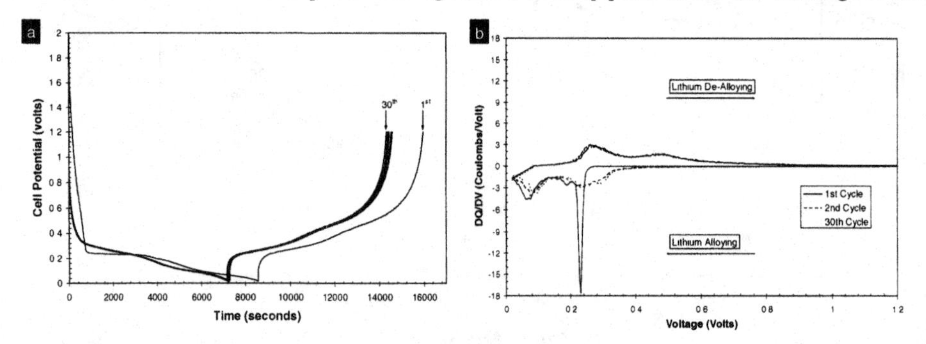

Figure 5. (a) Voltage versus time profile for 1st, 5th, 10th, 15th, 20th, 25th, and 30th cycles of 250 nm Si film on Cu foil (~C/2.5) and (b) corresponding calculated dQ/dV plots for 1st, 2nd, and 30th cycles.

differential capacity plot are unreacted silicon and Li_xSi (x~1.731, 2.367 for peaks at 0.26 V and 0.19 V, respectively)[27]. One can also note the depressed voltage at which the reactions begin (~262 mV vs Li counter-electrode) compared to the 332 mV observed at a higher temperature[27] (closer

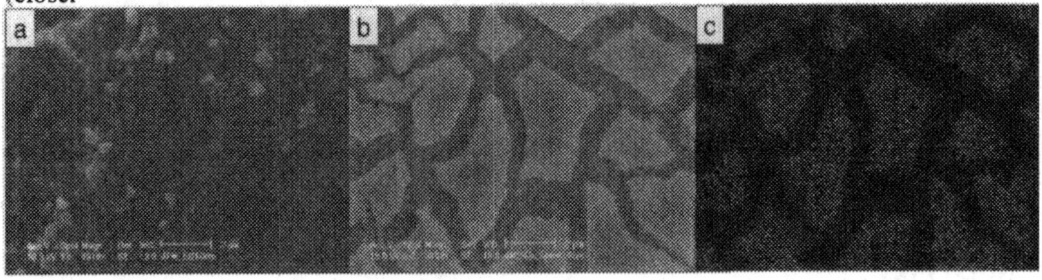

Figure 6. SEM morphology of (a) as-deposited 250 nm Si film on Cu, and (b) after 30th charge at C/2.5 rate. (c) Corresponding Si EDAX map taken on the same image area shown in part (b) (scale same in all images).

to thermodynamic equilibrium) for the onset of the Li-Si reaction. It is not immediately clear why the reaction onset occurs at a lower voltage. Polarization due to kinetic limitations of the as-deposited material could be suggested as a possible reason. The kinetics should depend on the initial density, the resistivity, and the lithium-ion diffusion coefficient in the material. Subsequent cycles however do not show the same depressed reaction potential and highlight the excellent reversibility of the 250 nm Si thin films. The subsequent cycles suggest the typical lithium alloying and de-alloying of an amorphous material with no two-phase regions present at any potential similar to that seen in amorphous $Si_{0.66}Sn_{0.34}$ reported by Beaulieu *et al.*[22] Impurities other than Ar in the Si film, such as O, H, CO, or N, if at all present, are expected to exist in only trace level. The base pressure of 5×10^{-7} Torr used in this study is in the range of the common base pressures used by researchers studying short range order in pure amorphous silicon.[28] Hence, no contamination can be expected.

The microstructure of the Si films both before and after cycling provides preliminary evidence about the possible mechanism responsible for the high reversible capacity retention seen in the 250 nm Si films. Fig. 6(a) shows the as-deposited 250 nm Si film on Cu. The film is fairly dense with ~200 nm diameter agglomerates on the surface. The surface agglomerates could possibly contribute to the irreversible capacity by separating from the film during the first cycle. Fig. 6(b) shows the morphology of the same film seen in Fig. 6(a) after completion of the 30th cycle charge. It can be seen that the silicon surface has cracked and separated into 1-3 μm diameter islands. In order to identify the composition of the cracked microstructures, EDAX was performed, the results of which are shown in Fig. 6(c). The EDAX analysis clearly shows that the cracked islands are essentially Si, while very little Si is seen in the valleys between the islands. Therefore, it appears that the islands of Si are formed during cycling directly on the Cu substrate, and not on an intermediate layer of silicon. The microstructure closely resembles that seen by Beaulieu *et al* for cycled Si-Sn films except that the active material islands seen in this work are in the size range of 1-4 μm that are about 15-20 times less than those seen by Beaulieu *et al.*[21] Further research is needed to identify the exact role of these islands of the active material formed on the copper substrate resulting in stabilization of the capacity. The mechanism may be similar to that observed by Beaulieu *et al*[21] in which the active material island has an outer shell region which reversibly expands and contracts during cycling, while a significant inner portion of the active material remains pinned to the Cu substrate providing an electronic pathway. More microstructural studies are needed however to verify these scientific aspects. The observed microstructural changes also suggest a possible explanation for the increase in capacity with cycle number for the 250 nm Si film cycled at 2C rate (Figure 4). It is possible that the high cycling rate precludes full lithiation through the complete depth of the film in the first cycle. Moreover, the high cycling rate creates a non-equilibrium condition affecting the kinetics of diffusion of the lithium ions on the surface of the cracked silicon. Therefore, a longer time is required for the lithium ions to completely alloy with silicon. Higher lithium incorporation could become possible in subsequent cycles due to cracking and "island" formation in surface layers, allowing uncycled silicon to become accessible to lithium ions from the electrolyte. Further microscopy work is in progress to completely understand the nature of the capacity increase with cycle number in the high rate cycled films.

CONCLUSION

Thin films of 250 nm Si sputter deposited on a Cu foil substrate yield steady reversible capacities close to 3500 mAh/g when tested for 30 cycles employing a C/2.5 current rate. Thicker 1.0 μm Si films exhibit steady reversible capacities of ~3000 mAh/g when tested at C/2.5 rate for 12 cycles, but fade substantially during subsequent cycles. Microstructural investigation of the as-deposited and post- 30 cycled 250 nm films revealed that the active material had formed Si islands on top of the Cu substrate which may reversibly expand and contract during cycling. The research performed in this study provides promise that a high capacity Si based anode material may one day replace carbon in commercial Li-ion batteries.

ACKNOWLEDGEMENTS

J.P.M. and P.N.K. acknowledge the support of NASA (GSRP and NAG3-2640), Chang's Ascending (Taiwan), NSF (CTS-0000563), and ONR(Grant N00014-00-1-0516). The authors also acknowledge useful technical discussions with Drs. George E. Blomgren and Michael H. Jin.

REFERENCES

[1] M. Winter and J.O. Besenhard, *Electrochim. Acta*, **45**, 31 (1999).

[2] Y. Idota, T. Kubota, A. Matsufuji, Y. Maekawa, and T. Miyasaka, *Science,* **276**, 1395 (1997).

[3] R.A. Huggins, *J. Power Sources,* **81-82**, 13 (1999).

[4] J.Yang, M. Winter, and J.O. Besenhard, *Solid State Ionics,* **90**, 281 (1996).

[5] R.A. Huggins, *Solid State Ionics*, **113-115**, 57 (1998).

[6] W.J. Weydanz, M. Wohlfahrt-Mehrens, and R.A. Huggins, *J. Power Sources*, **81-82**, 237 (1999).

[7] O. Mao, R. L. Turner, I. A. Courtney, B. D. Fredericksen, M. I. Buckett, L. J. Krause and J. R. Dahn, *Electrochem. Solid-State Lett.*, **2** (1), 3 (1999).

[8] A.M. Wilson, J.N. Reimers, E.W. Fuller, and J.R. Dahn, *Solid State lonics*, **74**, 249 (1994).

[9] W. Xing, A.M. Wilson, G. Zank, and J.R. Dahn, *Solid State Ionics*, **93**, 239 (1997).

[10] A.M. Wilson, W. Xing, G. Zank, B. Yates, and J.R. Dahn, *Solid State Ionics*, **100**, 259 (1997).

[11] I. Kim, P.N. Kumta, and G.E. Blomgren, *Electrochem. Solid-State Lett.*, **3** (11), 493 (2000).

[12] I. Kim, G.E. Blomgren, and P.N. Kumta, *Electrochem. Solid-State Lett.* **6** (8), A157 (2003).

[13] I. Kim, G.E. Blomgren, and P.N. Kumta, *J. Power Sources*, **130**, 275 (2004).

[14] I. Kim, G.E. Blomgren, and P.N. Kumta, *Electrochem. Solid-State Lett.*, **7** (3), A44 (2004).
[15] I. Kim and P.N. Kumta, *submitted to J. Power Sources* (2004).
[16] S. Song, K.A. Striebel, R.P. Reade, G.A. Roberts, and E.J. Cairns, *J. Electrochem. Soc.*, **150** (1), A121 (2003).
[17] J.P. Maranchi, A.F. Hepp, and P.N. Kumta, *Electrochem. Solid-State Lett.*, **6** (9), A198 (2003).
[18] M.N. Obrovac and L. Christensen, *Electrochem. Solid-State Lett.*, **7** (5), A93 (2004).
[19] S. Bourderau, T. Brousse, and D.M. Schleich, *J. Power Sources*, **81-82**, 233 (1999).
[20] S. Lee, J. Lee, S. Chung, H. Lee, S. Lee, and H. Baik, *J. Power Sources*, **97-98**, 191 (2001).
[21] L.Y. Beaulieu, K.W. Eberman, R.L. Turner, L.J. Krause, and J.R. Dahn, *Electrochem. Solid-State Lett.*, **4** (9), A137 (2001).
[22] L.Y. Beaulieu, K.C. Hewitt, R.L. Turner, A. Bonakdarpour, A.A. Abdo, L. Christensen, K.W. Eberman, L.J. Krause, and J.R. Dahn, *J. Electrochem. Soc.*, **150** (2), A149 (2003).
[23] K. Sayama, H. Yagi, Y. Kato, S. Matsuta, H. Tarui, and S. Fujitani, *Abstracts of the 11th International Meeting on Lithium Batteries*, 52, Monterey, CA (2002.) Available online at http://www.electrochem.org/meetings/satellite/imlb/11/abstracts/piimlb11.htm
[24] T. Yoshida, T. Fujihara, H. Fujimoto, R. Ohshita, M. Kamino, and S. Fujitani, *Abstracts of the 11th International Meeting on Lithium Batteries*, 48, Monterey, CA (2002.) Available online at http://www.electrochem.org/meetings/satellite/imlb/11/abstracts/piimlb11.htm
[25] T. Takamura, S. Ohara, J. Suzuki, and K. Sekine, *Abstracts of the 11th International Meeting on Lithium Batteries*, 257, Monterey, CA (2002.) Available online at http://www.electrochem.org/meetings/satellite/imlb/11/abstracts/piimlb11.htm
[26] I.C.D.D. Reference Pattern # 77-2109
[27] C.J. Wen and R.A. Huggins, *J. Solid State Chem.*, **37**, 271 (1981).
[28] P.M. Voyles, J.E. Gerbi, M.M.J. Treacy, J.M. Gibson, and J.R. Abelson, *J. Non-Cryst. Solids*, **293-295**, 45 (2001).

Chemically Derived Nano-encapsulated Tin-Carbon Composite Anodes for Li-ion Batteries

Il-seok Kim[a] and Prashant N. Kumta[a]

[a]Carnegie Mellon University, Pittsburgh, Pennsylvania 15213

G. E. Blomgren[a,*]

*Blomgren consulting Services Ltd., 1554 Clarence Avenue, Lakewood, Ohio 44107

ABSTRACT

Sn/C composites were synthesized by infiltrating different tin compounds into mechanically milled PS-resin powder. The composite obtained from tetraethyl tin and PS-resin powder followed by a heat-treatment of the precursor in Ar atmosphere at 600°C results in crystalline tin and amorphous carbon. The resultant composite exhibits a capacity of 480mAh/g with very good capacity retention (0.15 %loss/cycle). TEM analysis indicates nanocrystalline (~200nm) spherical tin particles embedded in large carbon particles (≅ 2 ~ 10 μm). Attainment of good cyclability appears to be governed by the microstructure of the composite.

INTRODUCTION

There has been extensive research effort to develop new anode systems that can outperform graphite systems currently used for Li-ion batteries. As a result, active/inactive nanocomposites containing electrochemically active Si and Sn have attracted significant interest due to their potential to generate high capacity.[1-6] Although those nanocomposites have shown their potential, no systems so far have attained any commercial status mainly due to the crucial problems related to the structural instability, caused by the large volume changes associated with alloying and dealloying of lithium during electrochemical cycling.[2-9] The large volumetric change resulting in cracking and crumbling of the electrodes causes abrupt failure of the electrode. Hence there is an imminent need for identifying systems and developing strategies for generating composites that can exhibit high reversible capacities

Our group has demonstrated the potential of Si/TiN, Si/TiB_2, and Si/SiC active/inactive nanocomposites for use as anodes in Li-ion batteries synthesized using high-energy mechanical milling (HEMM).[10-14] Despite these systems showing promise, exhibiting reasonably high capacity of 300~400mAh/g with stability, they present some limitations related to the loss in capacity after prolonged mechanical milling due to the synthetic methods used for generating the nanocomposites. Nanocomposites generated using HEMM exhibit significant capacity loss caused by the embedding of active species of Si during milling.[15]

In the present study, an alternative chemical synthetic method is presented to generated composite anode materials in order to overcome the capacity loss originating from HEMM. Tin has been chosen as a primary active species since Sn can be easily reduced chemically using common reducing agents such as alkali metals (Li and Na), alkaline-earth, hydrazine, sodium borohydride ($NaBH_4$) and even late transition metals such as Zn[16-18] in comparison to Si that cannot be easily synthesized using chemical synthetic methods. Similarly, C can be easily generated by thermal decomposition of various polymers[19] since C is the main non-volatile component of polymers generated by pyrolysis. Thus several researchers have studied the use of phenolic resins to generate carbonaceous materials for used as anodes,[20-22] and have even claimed that undesirable electrochemical properties such as irreversible loss can be controlled by appropriate heat-treatments. Based on the above analysis, carbon is justifiably used as the logical preferred matrix component since it is light and exhibits good electronic conductivity. Cross-linked polystyrene (PS)-based resin material has been used for the generation of the carbon matrix component since the polymer is capable of generating carbon at a reasonable yield (>35%) and will be able to accommodate the tin precursors that will ultimately yield Sn/C composite after decomposition induced by pyrolysis in inert atmospheres. Sn/C composites therefore were synthesized by infiltrating organometallic Sn sources into the PS-based resin followed by appropriate heat-treatments in the present work. Experimental studies and results of the electrochemical responses are thus presented in the present manuscript. In addition, microstructural analyses using transmission electron microscopy (TEM) have been conducted to identify the microstructural attributes necessary for attaining high capacity in this system.

EXPERIMENTAL

2g of poly(styrene) (PS) resin (Purolite Co. U.S.A., MN-200, Mw=100,000g/mole, >25% of Divinylbenzene, bead size:0.3~1.2mm) was mechanically milled (SPEX CertiPrep, 8000 M) for 30min to pulverize the bead-shaped particles into a fine powder. The PS-resin powder was then infiltrated using three alkyl tin compounds namely, as tetramethyl tin ($(CH_3)_4Sn$, Alfa Aesar, 98% assay), tetraethyl tin ($(C_2H_5)_4Sn$, Alfa Aesar, 98% assay) and tetra-n-octyl tin ($(C_8H_{17})_4Sn$, Gelest, 95% assay) with the nominal composition of Sn:C =1:1 at room temperature. In order to obtain complete infiltration, the precursor has been stirred using a stirring magnet for 1 h. The nominal composition of the precursor was Sn:C=1:1 based on the theoretical content of Sn in alkyl tin and the real weight of carbon residue obtained after pyrolysis of the PS-resin in ultra high purity (UHP) Ar. The as-prepared precursor was dried in a furnace for 2h at 150°C in UHP-Ar and then heat-treated subsequently at the desired temperature, 600 or 800 °C in the same furnace. The ramp rate was 300°C/hr and the flow rate of the UHP-Ar used was 100 ml/min.

Electrodes for electrochemical analysis were fabricated using the as-heat-treated powder by mixing 87.1 wt% of the active powder and 7.3 wt% acetylene black. A solution containing 5.6 wt% polyvinylidene fluoride (PVDF) in 1-methyl-2-pyrrolidinone (NMP) was added to the mixture. The as-prepared solution was coated onto a Cu foil (INSULECTRO, electro-deposited, thickness: 175μm). A hockey puck cell design was used employing lithium foil as an anode and 1M $LiPF_6$ in ethylene carbonate (EC) / dimethyl carbonate (DMC) (2:1) as the electrolyte. All the batteries tested in this study were cycled in the voltage range from 0.02~1.2V employing a current density of $0.1 mA/cm^2$ and a minute rest period between the charge/discharge cycles using a potentiostat (Arbin electrochemical instrument). The phases present in the heat-treated powder and the thermal decomposition behavior of the precursor were studied using x-ray diffraction (Philips X'PERT PRO system) and thermogravimetric analysis (TGA, TA Instruments, TA2960), while the microstructures of the precursor and the powder were examined using a scanning electron microscope (Philips XL30) and a transmission electron microscope (Philips EM420T).

RESULTS AND DISCUSSION

In order to generate the Sn/C composites, it is necessary to infiltrate the pores of the PS-resin, which does not dissolve in any known common solvents,[23] using a tin source that is preferably in a liquid state. PS-resin is available in the form of beads and the internal pores in the nanometer range within, allow a limited amount of liquid to be infiltrated into the pores. This unfortunately limits the total amount of tin that can be finally contained in the composite. In order to increase the amount of liquid that could be infiltrated into the PS-resin, fine sized PS-resin may be necessary, which could lead to an increase in the absorption of the liquid tin source on the surface due to the high surface area of the fine sized PS-resin powder. The SEM micrographs in Fig. 1 exhibit the morphologies of the PS-resin bead and powder obtained after mechanical milling for 30 min. The powder apparently has more open pore channels that were generated during milling by pulverizing beads into smaller particles. Therefore, the tin source is infiltrated into the powdered PS-resin, which is obtained after HEMM for

30 min. The BET surface area measurements show that the as-received PS-resin beads from the commercial vender and the powder obtained after mechanical milling for 30 min exhibit a specific surface area of 853 m^2/g and 961 m^2/g, respectively. These values indicate that the surface area of the PS-resin increases after mechanical milling. This extremely high surface area of the PS-resin powder appears to be ideal for conducting the infiltration experiments.

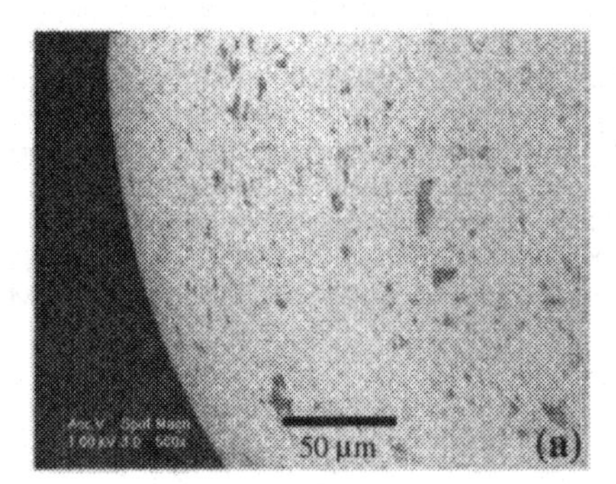

Figure 1. SEM micrographs showing the porous surface of (a) PS-resin bead and (b) PS-resin powder obtained after mechanically milling for 30 min.

Alkyl tin compounds have been studied for their ability to generate fine Sn for use as catalysts in the Pt-Sn system.[24, 25] They have therefore been used as the tin source for this research instead of tin chloride, the most commonly used tin source for generating elemental tin. Based on our preliminary study, all alkyl tin compounds decompose to generate elemental tin in the presence of carbon species. Hence several alkyl tin compounds have been selected to generate Sn/C composites. These include tetramethyl tin ($(CH_3)_4Sn$), tetraethyl tin ($(C_2H_5)_4Sn$), and tetra-n-octyl tin ($(C_8H_{17})_4Sn$). Two important aspects have been considered for the infiltration of alkyl tin into PS-resin. First, the molar volume of the alkyl tin compound should be low since it dictates the amount of tin that will be obtained in the final Sn/C composites obtained after pyrolysis. Second, the decomposition temperature is important, since it determines the temperature of formation of tin during heat-treatment. If the decomposition temperature is too low (< ~150°C) then the tin content in the composite may decrease during subsequent heating because of the evaporation of tin. However, the decomposition temperature and mechanisms involved in the pyrolysis of alkyl tin to yield metallic tin is not clearly understood at present. Nevertheless, in the present research, the boiling point has been used as the parameter for selecting the alkyl tin compound that will yield the optimum amount of tin in the final pyrolyzed composite. Since all alkyl tin compounds are liquids, no external solvent was used for conducting the infiltration experiments.

Fig. 2 exhibits the result of the XRD analyses of the Sn/C composites derived from various alkyl tin sources with a nominal Sn/C composition of Sn:C=1:1 after heat-treatment at 600°C in UHP-Ar for 5h. The nominal composition used for all these experiments has been determined using the theoretical Sn content in the alkyl tins and the yield of carbon obtained after pyrolysis of the PS-resin, since the PS-resin does not have a fixed chemical formula or composition. All the precursors were prepared by infiltration of each alkyl tin compound into the PS-resin powder obtained after milling for 30 min. The Sn/C composites were generated subsequently, after heat-treatment at 600°C in UHP-Ar for 5 h, wherein the PS-resin decomposes completely to form carbon. The XRD spectra of all the composites generated are very similar exhibiting sharp crystalline tin peaks while carbon does not show any essential peaks due to its amorphous nature.

The electrochemical responses of the Sn/C obtained from various alkyl tin compounds are shown in Fig. 3. The Sn/C composite derived from tetramethyl tin shows an initial discharge capacity higher than 300 mAh/g however, the capacity decreases slowly after 7 cycles (see Fig. 3 (a)). Although tetramethyl tin is advantageous for the infiltration of tin due to its low molar volume (120.63ml/mol), the Sn/C composite does not show high capacity. This appears to be attributed to loss of tin during subsequent pyrolysis after the decomposition of tetramethyl tin at a temperature as low as 74°C. The reason for the capacity fade is not clear at present, however particle growth may have caused the fade in capacity since tin particles form at an early stage of pyrolysis due to its low decomposition temperature. On the other hand, the composite derived from tetra-n-octyl tin shows a very low initial charge capacity of ~250 mAh/g (see Fig. 3 (b)), which appears to be caused by its high molar volume (389.54ml/mol). There is also significant fade in capacity possibly due to the presence of coarse-grained tin on the surface of the decomposed carbon, which undergoes cracking during cycling. Since there is limited volume of pores available in the PS-resin powder as mentioned before, the amount of tin obtained from infiltration of tetra-n-octyl tin might be negligible leading to the low capacity. Therefore it is our hypothesis that the alkyl tin compounds such as tetra-n-octyl tin which exhibit high molar volumes are not useful for generating Sn/C composite despite its high decomposition temperature (~224°C).

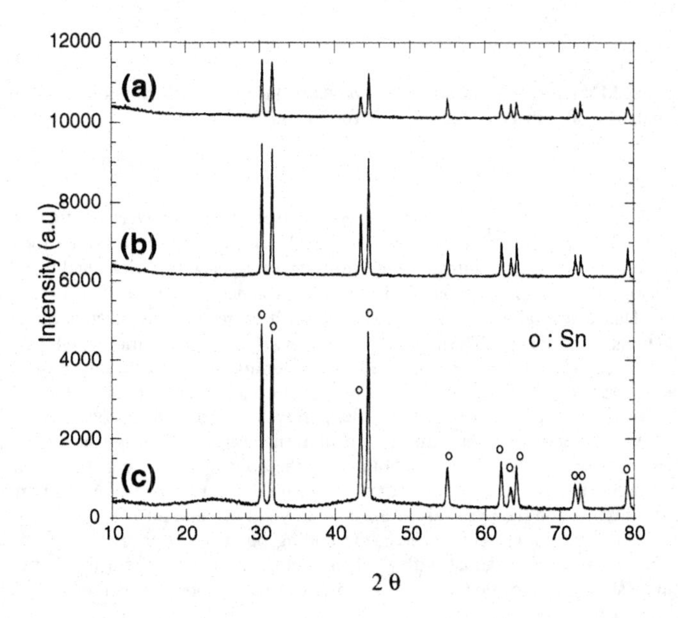

Figure 2. X-ray diffraction pattern of Sn/C composite corresponding to a composition of Sn:C=1:1obtained after infiltration of PS-resin with (a) tetramethyl tin, (b) tetra-n-octyl tin and (c) tetraethyl tin, followed by heat treatment for 5 h in UHP-Ar at 600°C.

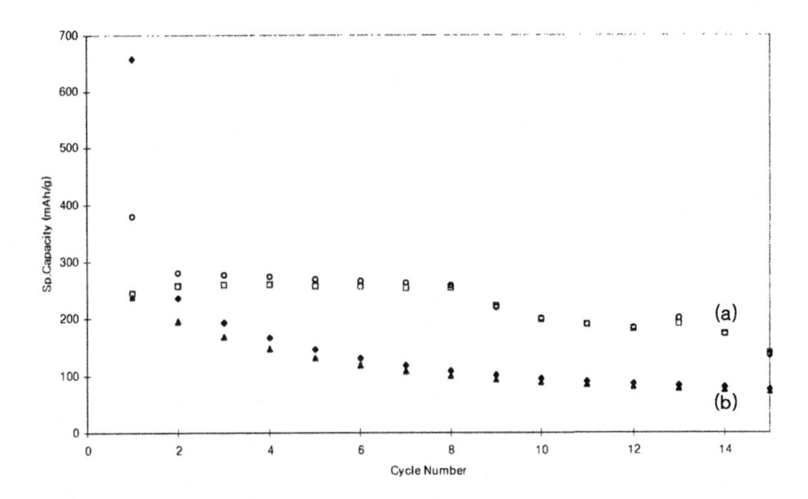

Figure 3. Capacity as a function of cycle number for Sn/C composites corresponding to a Sn:C =1:1 composition obtained after mechanical milling PS-resin powder infiltrated with (a) tetramethyl tin, and (b) tetra-n-octyl tin, respectively, followed by pyrolysis for 5 h in UHP-Ar at 600°C. (Current rate : 100μA/ cm^2, Potential : 0.02~1.2 V)

Based on the preliminary results obtained using different alkyl tin compounds, tetraethyl tin was selected as the tin precursor for infiltrating PS-resin to generate the Sn/C composites. The electrochemical response of this system is discussed below. The composite derived exhibits a relatively high capacity with stability compared to the other Sn/C composites derived from tetramethyl tin and tetra-n-octyl tin. Hence, alkyl tin compounds such as tetraethyl tin, which have a small molar volume (235.48ml/mol) and high boiling temperature (181°C) appear to be more appropriate for the purpose of infiltration of tin into the PS-resin powder due to the reasons mentioned previously.

Thermogravimetic analyses (TGA) of PS-resin and the final precursor obtained after infiltrating tetraethyl tin into the fine PS-resin powder have been conducted to investigate the decomposition of tetraethyl tin and PS-resin in UHP-Ar atmosphere using a heating rate 10°C/min and a flow rate of 100ml/min. As shown in Fig. 4 (b), there are two significant weight losses, seen at ~100°C and ~400°C, respectively. The first weight loss corresponds to the decomposition of tetraethyl tin and the second, is attributed to the decomposition of the PS-resin, which can be clearly seen in the thermal response of PS-resin only (Fig. 4 (a)). Although the boiling point of tetraethyl tin is 181°C, the decomposition appears to commence at a lower temperature. The nominal composition of Sn/C composite is Sn:C=1:1, calculated based on the molecular formula of tetraethyl tin. The first weight loss is attributed to the removal of four ethyl groups from the tetraethyl tin based on the TGA result of PS-resin powder, which does not show any weight loss up to 400 °C. The approximate actual composition of the Sn/C composite obtained after heat-treatment at 600°C could be estimated from the weight loss observed in the TGA analysis, to be Sn:C=1:5.56. This therefore indicates that the actual amount of tin in the composite is much less than the initial nominal composition (Sn:C=1:1) suggesting the loss of tin by evaporation during decomposition or subsequent heat-treatment.

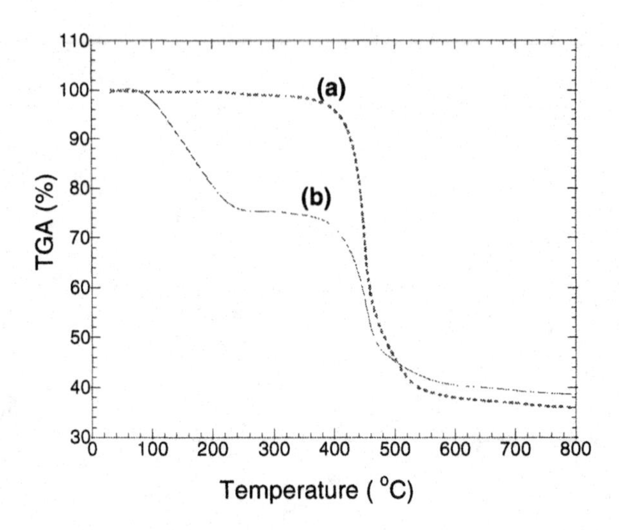

Figure 4. The TGA analysis of (a) PS-resin and (b) the precursor obtained after infiltrating tetraethyl tin into the PS-resin corresponding to the nominal composition of Sn:C=1:1 in UHP-Ar using a heating rate of 10°C/min.

In order to investigate the microstructure of the Sn/C composite obtained after heat-treatment for 5h at 600°C in UHP-Ar, TEM analysis was conducted on the composite and Fig. 5 shows the bright field image and the selected area diffraction pattern (SADP). The dark spherical regions in the BF image correspond to tin particles with a size of ~200 nm embedded inside a large carbon particle (≅ 2 ~ 10 μm). The SADP pattern corresponds to metallic tin. The microstructure shown in the micrograph can be used to speculate the mechanisms leading to the formation of the Sn/C composite during heat-treatment. It appears that tetraethyl tin decomposes first to form metallic tin, which then deposits on the surface of the PS-resin powder. At higher temperatures (>400°C) during pyrolysis, the PS-resin powder decomposes causing Sn to agglomerate and embed within the pyrolyzed carbon to generate the specific microstructure shown in Fig. 5 (a). Since Sn is embedded within the thick carbon particle, it is very difficult to distinctly image the tin particles. However, the SADP (Fig. 5 (b)) consisting of spots and diffused halos clearly suggest that crystalline tin particles coexist within the amorphous carbon matrix. The SADP analysis therefore validates the XRD results.

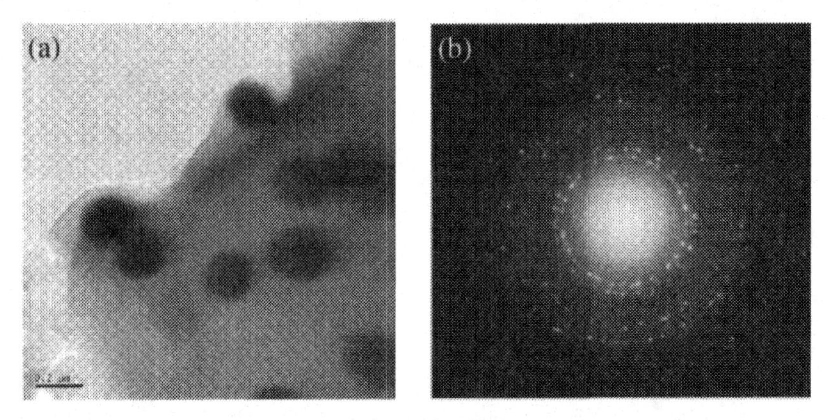

Figure 5. TEM micrographs of the Sn/C composite obtained from PS-resin infiltrated with tetraethyl tin corresponding to a composition of Sn:C=1:1 after heat treatment for 5 h in UHP-Ar at 600°C; (a) BF image, (b) SA diffraction pattern (camera length = 61cm, reduced to 40% of its original size).

The Sn/C composite derived from the PS-resin powder infiltrated with tetraethyl tin after heat-treatment at 600°C for 5h in UHP-Ar has been analyzed for its electrochemical properties using a current rate of 100 $\mu A/cm^2$ (≅ C-rate of C/13). The capacity vs. cycle number of Sn/C composites is shown in Fig. 6 (a). The nominal composition used for synthesizing the composite corresponds to Sn:C=1:1, although the actual composition is 1:5.56 determined from the TGA result as discussed above. Based on the actual composition, the theoretical capacity of the composite should be 636mAh/g considering that the capacity is only associated with tin in the composite, which is slightly lower than the first discharge capacity (~718 mAh/g) shown in the plot. However the carbon derived from PS-resin alone pyrolyzed at 600 °C for 5 h in UHP-Ar also exhibits an initial discharge capacity of 750 mAh/g although a stable capacity ~300 mAh/g is ultimately realized according to our preliminary results. Thus the overall capacity is likely to originate from carbon as well as tin. It can be mentioned that the capacity of the Sn/C composite derived from tetraethyl tin is less than the sum of the theoretical capacity and the capacity contributed by carbon although the contribution from carbon cannot be determined exactly. The reason for the discrepancy between the actual and the estimated capacity is not clear at this moment, however partial embedding of Sn inside the large carbon particles appears to be one of the factors contributing to this difference. The embedded tin is surrounded by carbon probably preventing the lithium ions from completely reacting with tin. More detailed impedance analysis is warranted to validate this hypothesis. The first irreversible loss of the composite (indicated by the peak appearing at ~ 0.14 V in the first discharge of the differential capacity plot shown in Fig. 6b) appears to be caused by carbon since a similar trend is seen when the pyrolyzed carbon obtained from PS-resin alone is electrochemically tested, results of which are not shown. The irreversible loss in carbon materials may be caused by the disordered carbon structure of amorphous carbons[22] or the presence of hydrogen attached to carbon that is also common in carbon derived from polymeric resins.[26, 27]

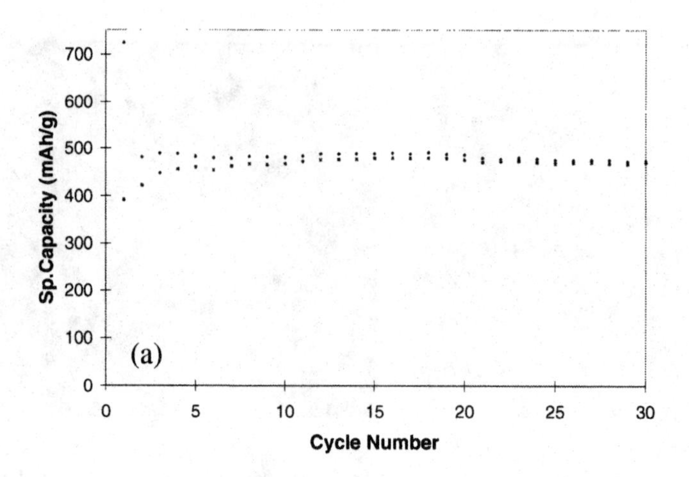

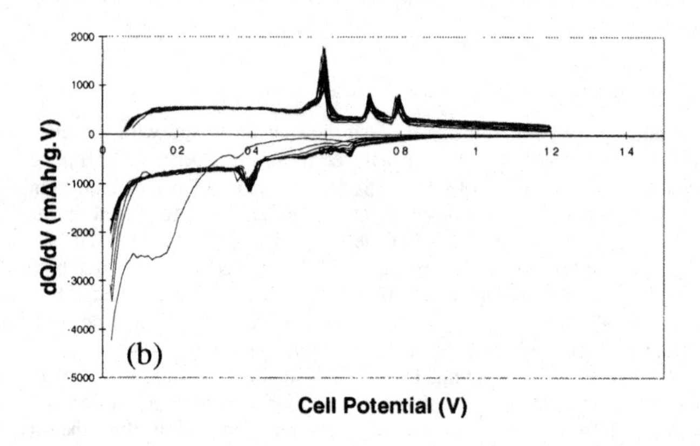

Figure 6. (a) Plot of capacity as a function of cycle number, (b) Differential capacity vs. cell potential curves obtained after infiltration of PS-resin with tetraethyl tin corresponding to the composition of Sn:C=1:1 following heat treatment for 5 h in UHP-Ar at 600 °C (Current rate : 100 μA/ cm^2, Potential : 0.02~1.2 V).

The Sn/C composite electrode exhibits an initial gravimetric capacity of 490 mAh/g during second discharge and the first charge capacity of ~400 mAh/g. However, with subsequent cycles there is an increase in the coulombic efficiency leading to a stable capacity of ~480 mAh/g, which is equivalent to a volumetric capacity of ~1450 Ah/l, at the end of the 30th cycle. The composite also displays good capacity retention (0.15%loss/cycle). This volumetric capacity is significantly higher (≅ 75%) than that of carbon as indicated in the plot. This observation supports the TEM result, that the Sn particles are homogeneously dispersed within the carbon particle, which leads to good capacity retention. However, partial embedment of tin within the carbon matrix may also contribute to the observed lower discharge capacity compared to its theoretical capacity.

Fig. 6 (b) shows the differential capacity plot of the Sn/C composites obtained after heat-treatment at 600°C for 5h in UHP-Ar. The sharp peaks during charge and discharge processes are attributed to the nanocrystalline tin in the composite. All of the peaks can be attributed to the presence of Li_xSn phases, where $0<x<4.4$.[1] Although there are no peaks related to carbon, the background is caused by carbon indicating its contribution to the overall capacity. The peak intensity does not change during cycling, which suggests the good cyclability of the composite. It can thus be concluded based on this result that the Sn/C composite prepared from tetraethyl tin and PS-resin powder is very promising for use as anodes in Li-ion application and there is clearly a need for further optimization. There are also key fundamental questions with regards to the decomposition mechanisms, the structural evolution of the composite and their combined influence on the electrochemical properties. These are all aspects demanding further research to be conducted in this promising system and the infiltration approach developed in the present study.

CONCLUSIONS

Nanostructured Sn/C composites can be produced by infiltrating alkyl tin compounds into mechanically milled PS-resin powder, followed by heat treatments in UHP-Ar atmosphere. The amount of tin remaining in the composites obtained from different tin sources after pyrolysis appears to be closely related to the molar volume and the boiling temperature of the alkyl tin compounds. The XRD analysis conducted on the Sn/C composite derived from infiltration of tetraethyl tin into the PS-resin powder after pyrolysis at 600°C for 5h in UHP-Ar indicates that the composite is composed of amorphous carbon and crystalline Sn. TEM analysis of the Sn/C composite obtained from tetraethyl tin and PS-resin indicate that spherical, nanocrystalline (~200nm) tin particles are largely embedded in the large carbon particles (≅ 2 ~ 10 μm). The Sn/C composite shows a capacity of ~480mAh/g with very good capacity retention (0.15 %loss/cycle), suggesting the promising nature of this approach for Li-ion application.

ACKNOWLEDGEMENTS

Il-Seok Kim, P. N. Kumta and G. E. Blomgren would like to thank the financial support of NSF (Grants CTS-9700343 and CTS-0000563), NASA (NAG3-2640) and ONR (Grant N00014-00-1-0516) for this work.

REFERENCES

1. M. Winter and J. O. Besenhard, *Electrochim. Acta*, **45**, 31 (1999).
2. O. Mao, R. L. Turner, I. A. Courtney, B. D. Fredericksen, M. I. Buckett, L. J. Krause and J. R. Dahn, *Electrochem. Solid-State Lett.*, **2**, 3 (1999).
3. O. Mao and J. R. Dahn, *J. Electrochem. Soc.*, **146**, 405 (1999).
4. J. Yang, , M. Wachtler, M. Winters and J. O. Besenhard, *Electrochem. Solid-State Lett.*, **2**, 161 (1999).
5. G. M. Ehrlich, C. Durand, X. Chen, T. A. Hugener, F. Spiess and S. L. Suib, *J. Electrochem. Soc.*, **147**, 886 (2000).
6. H. Kim, B. Park, H. Sohn and T. Kang, *J. Power Sources*, **90**, 59 (2000).
7. J. Niu and J. Y. Lee, *Electrochem. Solid-State Lett.*, **5**, A107 (2002).
8. M. Egashira, H. Takatsuji, S. Okada and J-i. Yamaki, *J. Power Sources*, **107**, 56 (2002).
9. L. Shi, H. Li, Z. Wang, X. Huang and L. Chen, *J. Mater. Chem.*, **11**, 1502 (2001).
10. I. S. Kim, G. E. Blomgren and P. N. Kumta, *Electrochem. Solid-State Lett.*, **3**, 493 (2000).
11. I. S. Kim, G. E. Blomgren and P. N. Kumta, *Ceramic Transactions*, **249**, 35 (2002).

12. I. S. Kim, G. E. Blomgren and P. N. *Kumta, Ceramic Transactions*, **249**, 127 (2002).
13. I. S. Kim, G. E. Blomgren and P. N. Kumta, *Electrochem. Solid-State Lett.*, **6**, A157 (2003).
14. I. S. Kim, G. E. Blomgren and P. N. Kumta, *Electrochem. Solid-State Lett.*, **7**, A44 (2003).
15. I. S. Kim, Ph. D. thesis, 'Synthesis, structure and properties of electrochemically active nanocomposites' (2003).
16. M. Christianson, D. Price and R. Whitehead, *J. Organomet. Chem.*, **102**, 273 (1975).
17. E. Borsella, A. Nesterenko and R. Larciprete, *Chem. Phys. Lett.*, **199**, 605 (1992).
18. L. Shi, H. Li, Z. Wang X. Huang and L. Chen, *J. Mater. Chem.*, **11**, 1502 (2001).
19. S. E. Hayes, H. Eckert and W. R. Even Jr, *J. Electrochem. Soc.*, **146**, 2435 (1999).
20. Z. Wang, X. Huang and L. Chen, *J. Power Sources*, **81-82**, 328 (1999).
21. T. Zheng, Q. Zhong and J. R. Dahn, *J. Electrochem. Soc.*, **142**, L211 (1995).
22. M. Noel and V. Suryanarayanan, *J. Power Sources*, **111**, 193 (2002).
23. Technical Data, The Purolite Co., Bala Cynwyd, PA 19004.
24. J. L. Margitfalvi, I. Kolosova, E. Talas and S. Gobolos, *Appl. Catal.*, A **154,** L1 (1997).
25. J. L. Margitfalvi, I. Borbath, M. Hegedus and A. Tompos, *Appl. Catal.*, A **229**, 35 (2002).
26. T. Zheng, Q. Zhong and J. R. Dahn, *J. Electrochem. Soc.*, **142**, L211 (1995).
27. Y. Liu, J. S. Xue, T. Zheng and J. R. Dahn, *Carbon* **34**, 193 (1996).

Author Index

Keyword Index

9 781574 981827